Carnivore Minds

Carnivore Minds

Who These Fearsome Animals Really Are

G. A. Bradshaw

Yale UNIVERSITY PRESS NEW HAVEN & LONDON

Frontispiece: Marten with remains of Elk. Photo credit: Charlie Russell. *Figure on page xix:* Photo credit: Charlie Russell and Paul Zakora. *Figure on page 1:* Photo credit: Fred Buyle.

Published with assistance from the Louis Stern Memorial Fund.

Yale University Press books may be purchased in quantity for educational, business, or promotional use. For information, please e-mail sales.press@yale.edu (U.S. office) or sales@yaleup.co.uk (U.K. office).

Set in Adobe Garamond and Scala Sans type by
Integrated Publishing Solutions.
Printed in the United States of America.

ISBN 978-0-300-21815-2 (hardcover : alk. paper)

Library of Congress Control Number: 2016952232

A catalogue record for this book is available from the British Library.

This paper meets the requirements of ANSI/NISO Z39.48–1992 (Permanence of Paper).

10 9 8 7 6 5 4 3 2 1

RRB, SBB, PBB, AGS, BRT, TET, TVT

Mary Watkins, an unbounded beacon

and always, JGB

Contents

Giving Voice to Animals

A Naturalist's Note

You should brace yourself for a measure of truth when you pick up this book. For a long time, the truth has been unwelcome and telling it, if not "a revolutionary act," is, at the very least, a thankless job. We are usually navigated away from facts and then told what to think; we are managed as surely as wildlife are managed, by those Gay Bradshaw calls the privileged authority. I do not like the feeling of being managed and have always wanted to understand what is really going on in what has become a human-controlled world. Or go back and live again deep in the heart of nature, but ironically I have become dependent on a drug that keeps me alive. Because of certain health issues, I am now part of the human experiment that is based on superseding nature. Although I fear that the whole experiment is bound to fail, I would have died long ago without all sorts of clever interventions. I am thankful for my extended years, but at the same time realistic about how precarious our entire manipulated survival system is.

I have always had a home base on the land that my grandfather homesteaded 110 years ago, next to a national park in the foothills of the Alberta Rockies. Because most of my life has been spent in the same place, I have witnessed a remarkable cycle. When I was a kid, there were no wolves and virtually no coyotes, foxes, cougars, or martins, and very few bears, eagles, or hawks. I thought it was normal to be swimming among schools of white fish in the river, tripping over rabbits in the bush, finding voles under every hay bale which were chewing through the strings, and elk eating most of the winter haystack before we could feed it to the cows. There was a plentitude of herbivores, because virtually all predators had been trapped, shot, or poisoned. That was our lifestyle in those days, even in the national parks. Deer and grouse and other "good-to-eat" fauna were "okay animals"—bears and other wildlife with teeth and claws were not "okay," because they were perceived as competition for humans.

Over time, this lopsided abundance began to disappear. At first, I thought that something was very wrong, and I blamed it on phenomena such as acid rain or other unintended consequences of the industrialized world. But, gradually, I realized that it was because the predators were coming back. The fur trade was losing traction, and city people from around the world were flocking to the parks to see predators—the bears, wolves, and osprey. All of a sudden *dangerous* animals had become *attractive* animals, because they were going extinct.

Renewed appreciation for predators plays an important role in protecting wildlife who have never been afforded protection. Nonetheless, even though public opinion has changed, and even though what science knows has changed, official rhetoric has been very slow to keep up. Park and public land managers still villainize predators. Yet everything that is said about predator danger—that grizzlies are unpredictable, aggressive, and become more dangerous unless they are treated with hostility—is completely at odds with my experiences from seven decades of living with them up close, on a daily basis.

As long ago as the 1960s, I recognized that grizzly bears were being talked about in ways that did not make sense to me. I had limited opportunities to watch them when I was young, even though I had been raised among the few who were left after the ranchers and hunters slaughtered them without question. I began to learn more deeply about bears when my father took on a project to make the first documentary on how grizzlies lived in the wild. We had to film it in the few places left in North America, mostly in Alaska and British Columbia, where there were still grizzlies living a normal life. Disney was making films about predators, including grizzlies, who were wild, but really all the filming was done with enclosed, controlled animals.

Being my father's camera man for three years in my twenties and wandering among grizzlies in the beauty of their habitat, I was able to observe them closely. For some reason I had a strong interest in understanding their true nature. I seemed to be able to identify with them, and what I saw vanquished any worries for my personal safety. I was able to get rid of much of the fear that had been instilled by endless stories of violence and mayhem that accompany almost every mention of grizzly bears. After spending most of my life exploring how to be able to trust grizzly bears, I am convinced that we can trust these bears implicitly. But there are far too many reasons why bears cannot, and should not, trust us.

Much later, when I was living with hundreds of bears in the wild, pristine nature of Russia for an amazing ten years, it became clear that I did not want to leave the wilderness. The world of the grizzly is a wonderful world and it lacks any of the burdensome problems of humanity. Yet I did have to venture back into civilization periodically, and the more I learned about bears, the more I wanted to find a way to apply what I had learned and put it toward something useful.

Once I understood how the real bear wanted to live, I wanted to make it plain that these animals are peace-loving, sensitive, and very intelligent, not the popular image of a blood-thirsty, horrible, unpredictable monster. It puzzled me for many years why this myth persists. Then I discovered that a lot of bad press about bears is generated by the hunting industry, which supports government wildlife agencies. The more hype about how scary an animal like a bear or wolf supposedly is, the more money can be charged for the opportunity to kill it. It is a legacy of how the system was set up in the early days of pioneering. As a result, a whole culture of fear has been built up around grizzlies.

I am excited about this book for two reasons. First, the book says in straightforward science what I have observed in Russia and Canada living among grizzlies, cougars, and other predators. Second, *Carnivore Minds* answers any remaining questions that I have had. Backed by science, Gay Bradshaw describes the true nature of predators: sentient kin who show what it means to live ethically and peacefully. Having Gay explain what is happening from the standpoint of science has also helped me sort out opinions from reality. She is a genius at figuring out how to explain phenomena credibly and clearly, so most people can understand why these animals behave as they do. I am ever grateful because, before she came along, all my observations were regarded by wildlife managers and a lot of scientists as anecdotal.

This book gives us great hope for the future. The plight of the bear and other carnivores is so dire that I was getting close to thinking that all of my hard work was not usable. But then when I became friends with Gay, I found out that what I have learned is in agreement with science after all.

What most strikes me about Gay and her work and what makes her stand out is that I saw that she was telling the truth. In that way she was just like another one of my bears and very unlike most other scientists. Using the careful logic of science, Gay painstakingly and eloquently rights the record. After five and a half decades of my own struggling to convince the world how wonderful bears really are, Gay says it all in the language of science and psychology. It is wonderful to finally see bears and all the other misrepresented and misunderstood animals have a chance to speak their truth. This book gives the animals a voice.

Charlie Russell

Foreword

Gay Bradshaw's goal in writing this book is extremely ambitious—to attempt to synthesize the biological, neuroscientific, psychological, and psychiatric literatures to provide a deeper understanding of the behavior, brains, and minds of carnivores. That said, she is uniquely qualified to write this volume. Over the course of her prolific career she has studied and worked intimately with a diverse array of species as an ethologist, psychologist, and indeed, psychotherapist of animals. In addition to possessing a remarkable breadth of experience, she is extraordinarily adept at not only reporting very recent advances within a number of different scientific literatures, but also creatively bridging the data and forging conceptual links across these bodies of knowledge. Working at the cutting edge and the interface of disciplines, this groundbreaking book is an exceptional feat of scholarship. In this labor of love Bradshaw offers the reader the wealth of her own experience—near observations of the social and emotional behaviors of a number of different animals in the wild—along with numerous reports from the world's leading biologists who have also worked closely with various carnivore species. Even more, she presents a compelling theoretical integration of biology and psychology to model the deeper brain-mind mechanisms that underlie the adaptive functions that have contributed to the survival of both carnivores and humans.

The keystone of Bradshaw's model lies in the synthesis of a rapidly expanding body of very recent research on the neurobiology of social and emotional abilities in both humans and animals. Indeed this interdisciplinary perspective characterizes her earlier studies of elephants, who share very specific "higher" functions that are admired by humans. But here she leaps from the mind of a gentle giant into the minds of various carnivores who are most feared by humans. In the following chapters, she offers numerous evocative portraits of the cognitive and particularly social and emotional functions of the carnivore's mind. To shift the focus of scientific inquiry from outer behavior to a deeper understanding of the inner world of carnivores, Bradshaw offers a neurological lens through which to view their natural histories, and so brings to light "neuroscience's long-standing, species-common model of brain, mind, and consciousness." Note the bold leap from the science of the behavior of individual animals in natural physical environments to the science of interacting animals, emotion, and consciousness in social environments.

The linchpin of Bradshaw's model is the integration of biology and neuroscience to generate a greater apprehension of the common adaptive (and maladaptive) social and emotional mechanisms that are shared by humans and all animals, including carnivores. In particular, at various points Bradshaw describes ongoing research in humans on brain lateralization and the functional asymmetry of the two cerebral hemispheres. She specifically refers to research in humans on the critical role of right lateralized cortical-subcortical systems involved in processing social and emotional information, and to the "mammalian right hemisphere's socioaffective centers—the areas key to stress regulation, self-development, and affiliative behavior." Here I'd like to point out that there already exists a sizable literature on brain and behavioral lateralization in animals. For much of the last two centuries science held the erroneous assumption that because only humans possess left brain specializations for language, brain lateralization was a uniquely human function. With the onset of this century, however, brain laterality research in animals has exploded. It can now be used to offer important insights into the minds of carnivores, and indeed into common nonverbal psychoneurobiological survival mechanisms of not only mammals and humans, but also all vertebrates.

Indeed, two recent influential comprehensive reviews cite a large body of research that documents the lateralization of specific functions of the right and left brains in both domesticated and wild animals in their natural habitats.[1] This burgeoning transspecies literature on hemispheric dominance also reports structural and functional differences in the two hemispheres of the human brain, showing continuity of these lateralized functions across the phylogeny of species. It is now thought that the evolutionary mechanism of brain asymmetry in vertebrates emerged about five hundred million years ago. From this point of origin the right hemisphere, the primary seat of emotional arousal, is believed to have become specialized for detecting and responding to unexpected stimuli in the environment and for controlling escape and other emergency responses, in contrast to the left brain, which originally became specialized for controlling well-established patterns of behavior under ordinary and familiar circumstances. To sustain these lateralized functions, the right hemisphere utilizes a global form of wide-ranging attention to attend to novelty, while the left uses a narrowly focused attention. For example, there is now agreement that in fishes, amphibians, reptiles, birds, and mammals the right hemisphere is specialized to detect and rapidly escape from carnivores while the left hemisphere is specialized for "prey" capture, a finding directly relevant to the theme of this book. These studies demonstrate that across species, animals keep a watchful left eye on any visible carnivores, respond more strongly to a potential carnivore seen on their left side, and react with greater avoidance to carnivores seen in the left side of their visual field (right side of the brain) than in their right visual field.

Another shared major tenet of both animal and human brain lateralization research is that the development of hemispheric specialization is strongly influenced by social attachment experiences and reproductive hormones in early life, and that stimulation of the developing young leads to profound and long-lasting alterations in hemispheric control, especially in the regulation of emotions. Supporting Darwin's pioneering work on the expression of emotions in man and animals, it is now held that "the right hemisphere is dominant in the control of strong emotions, especially hostility and aggression, as demonstrated by increased levels of blood flow and neural activity in regions of the right hemisphere when these emotions are expressed."[2] In addition, both literatures document that early social context indelibly influences the most fundamental structural system involved in stress regulation—the hypothalamic-pituitary-adrenal axis—and that for the rest of the lifespan the right hemisphere controls physiological stress responses and cortisol production. Parallel with this discovery are those from animal and human studies clearly indicating that the long-term effects of early traumatic social stress include a shift to a cognitive-emotional bias toward a predominant negative state and an enduring alteration of right brain functions, expressed in future impaired abilities for coping with social stress and a predisposition to post-traumatic stress disorder.

In sum, then, the discussions of nonhuman animals in this book are congruent with human models of interpersonal neurobiology, an interdisciplinary perspective that focuses on how early social-emotional attachment experiences, for better or worse, shape the developing brain/mind systems. Bradshaw's earlier application of human interpersonal neurobiological models of development to anthropogenically stressed elephant populations established the validity and utility of understanding the animal psychopathology of "brutal killing and sexual assaults of rhinoceroses" in terms of early relational trauma-induced post-traumatic stress disorder. This novel conceptualization of the origins and expression of aberrant animal behavior in terms of these human models shifted the paradigm for wildlife biology.

Equally ground-breaking, Bradshaw masterfully provides evocative and compelling insights into the minds of a spectrum of carnivores, attempting to counter the extant demonization and destruction of these species. Although anthropomorphism was viewed as an antiscientific principle, humans have a long history of freely projecting their worst fears and aggressive tendencies onto the animals in this book. In addition to imparting a deeper understanding of the consciousness of animal minds, then, Bradshaw brings to light the "vanquishing myths that mask the true identity of carnivores," and offers a plea for the ethical responsibility of our species to the welfare of other species that co-inhabit the planet. Bradshaw calls for not only a deeper understanding of the minds of carnivores and wildlife in general, but also a new paradigm that "demands a change in how the world is perceived and . . . a change in how we ourselves are viewed."

Ultimately, this change entails more than coming to a deeper understanding of the consciousness of carnivores: it requires investigating various states of consciousness of humans, the apex carnivore of the planet and the disrupter of free-ranging populations of carnivores. This kind of expanded self-inquiry must include both the reflected objective awareness of the human surface, conscious mind located in the left brain, as well as the reflective empathic social-emotional awareness of the deeper unconscious mind located in the right brain.

As the reader will see, the creative narrative style of this book includes frequent descriptions of social-emotional communications between animal minds and bodies. Offering more than an abstract, left-hemispheric, objective, and "coldly scientific" discussion of animal behavior, Bradshaw writes in a right-hemispheric, emotionally evocative, imagistic way, describing not just the behavior but also the carnivores' subjective states of mind. By doing so, she moves them away from cold, lethal abstractions, helping us to understand them as sentient living beings with intimate social needs and emotional and motivational systems that are in some very basic ways similar to those of humans.

Indeed, as a colleague and coauthor I can attest to Gay Bradshaw's remarkable abilities as a scientist to fully engage her right brain creativity and curiosity to make intersubjective contact with the emotional subjectivities of a wide variety of animal species. Going beyond the now established principle that animals, like humans, have adaptive emotions and engage in intensely social lives, she uses her own intuitive, bodily based emotional reactions to read the external outputs and internal states of mind of the emotional brains of animals. This orientation is also central to her pioneering studies on the psychobiological treatment of traumatized animals with methods that access right brain empathy, a central mechanism of change in the human psychotherapy literature.[3] The studies she reports here of her own work as well as that of other biologists studying carnivores in an experience-near fashion are establishing not only the importance of research on the interpersonal neurobiology of intraspecies social-emotional communications, but also the validity of interspecies communication as a scientific method in behavioral biology.

As Bradshaw repeatedly demonstrates, to develop a more comprehensive understanding of the natural lives of carnivores, biologists need to move from the stance of detached observer to that of participant observer, and to read their subjects' intimate right-brain-to-right-brain communications when they are in a state of safety with a human being. The formal study of animals within their relational contexts not only enriches our scientific understanding of nature, but also can offer invaluable and deeply rewarding emotional experiences and expansions of human consciousness (see the remarkably evocative, inspiring description of human contact with sperm whales in the Epilogue).

But as the reader will soon see, the following pages contain more than uplifting, positive glimpses into the psyches and lives of the "toothed and clawed." Bradshaw vividly describes not only her own observations but also those of other scientists who are documenting the rapid anthropogenic decline of the "wild" species described in this book. These disturbing images of humans' destruction of the biota are strongly reminiscent of Rachel Carson's evocative descriptions of a changing botanical landscape blighted by the unrestricted proliferation of chemical pesticides, another form of anthropogenic disturbance. Her transformational *Silent Spring* was a call for action that compelled changes in governmental regulation. In this similarly trailblazing volume Bradshaw urgently and passionately sounds a similar clarion call for political action and moral change in our relationship to not only carnivores but also the natural world.

This pioneering work makes important and indeed essential contributions to ethology, the branch of science that deals with animal behavior, especially in the wild, as well as to conservation biology, the branch of biology that deals with threats to biodiversity and with preserving the biologic and genetic diversity of animals and plants. It is therefore essential reading for not only researchers, conservationists and policy makers, but for anyone who cares about the welfare of life on our planet. It is my hope and prediction that this extraordinary work will become a classic and transformational volume in both the biological and psychological literatures.

Allan N. Schore

Preface

The inspiration for this book is the enigmatic grizzly bear. A few years back, Charlie Russell and I started talking about bears and what he called "the mess that we humans have gotten bears into." He had read *Elephants on the Edge* and I, *Grizzly Hearts.* Together, we discovered common ground at the intersection of my science with his lifetime of experience. One conversation turned into two, then very quickly, into weekly meetings during which we tossed back and forth our stories, ideas, and frustrations.

We had a lot to talk about because the animal species that we were trying so passionately to save have much in common. Bears, like elephants, were under siege. Bears, like elephants, had lost the land and livelihoods that had sustained them for thousands of years. Bears and elephants were acting un-bear- and un-elephant-like and it did not take any abstract theories to figure out why. Both species had reached a breaking point after centuries of unrelenting human violence. But although bears and elephants have had similar fates, it wasn't easy to convince people of the symmetry.

Elephant PTSD (post-traumatic stress disorder) was obvious. The pieces of the trauma puzzle fit together with tragic ease. People intuitively accept that brutal killing and sexual assaults of rhinoceroses by the gentle giants are symptoms of profound distress. It wasn't normal for the otherwise pacific, family-oriented elephants to show anything but love and compassion for their own species and for others. It made sense that highly sensitive pachyderms would crack under the grief and shock of seeing their families gunned down, ivory tusks ripped off still-breathing bodies. Making the case for grizzlies was not so clear, nor was their plight so readily appreciated.

Unlike the herbivorous elephant, grizzles are members of the *Carnivora* order, animals who eat other animals. But this classification is somewhat misleading. In contrast to obligate carnivores, such as pumas, white sharks, and rattlesnakes, grizzlies are largely plant-based eaters who feast on grass, roots, nuts, and berries. The salmon, occasional rodent, and larvae that they consume, while providing a vital infusion of calories necessary to withstand winter hibernation, constitute only a portion of the bears' total diet. Nonetheless, the grizzly is grouped with full-time flesh eaters as a fellow predator.

While it usually implies carnivore, "predator" speaks to another quality that is attributed to grizzlies. Bears, crocodiles, and sharks are considered killers whose prime directive is the use of violence to solve problems, whether those problems concern food, mates, or territory. Most scientists will concede that an African lion pride forms the nucleus of some sort of society, but pumas and grizzlies are viewed as infamous loners more likely to slay each other than to enjoy any alliance. Nor are these species known, as are elephants, for touchy-feely ways with their offspring. Instead, the carnivore image is more like a furred or finned psychopath who not infrequently commits infanticide. Given all this, how could anyone tell if a grizzly exhibited symptoms of PTSD, let alone expect that bears, being agents of bloodshed themselves, would be susceptible to psychological trauma? Right? Wrong.

It didn't take long for Charlie to disabuse popular assumptions of bear unpredictability, unprovoked aggression, and danger in proximity to humans. His seventy-odd years of survival living in the Canadian Rockies and Siberian cold, often literally side by side with grizzlies, belied the myths and rang true with what neuroscience predicts: grizzlies and all other vertebrate species share and exhibit prosocial mental and emotional capacities that are governed by brain structures and processes common to elephants and humans. With Charlie's encouragement, and my belief that an integrated analysis of the science would reconcile conflicting views, I began to investigate the nature of these magical, mystical, supposedly monstrous beasts by approaching their natural histories through a neuropsychological lens.

The marriage of wildlife biology and neuroscience is still new, and for the most part, unconsummated. When there has been a joining of disciplines, such as, for example, the routine use of monkeys in studies of human psychopathology or experiments on elephant reactions to LSD, it has usually brought a world of hurt to animals. Cats, rats, chimpanzees, pigeons, beagles, rabbits, octopi, planarians, and species from other taxonomic branches in the tree of life have been used as human surrogates in research prohibited for our species, yet denied even commonsense protections that any living being—"surrogate" or not—deserves. While science acknowledges human-nonhuman equivalence in concept (theory) and practice (experimentation), neuroscientific symmetry fails to translate into ethical parity. Furthermore, what has been learned about animal minds in the laboratory has not been incorporated into animal studies in the field.

It took the rash of unprecedented, abnormal behavior of free-living African elephants to break through the conceptual dam that has held back the natural flow of communication between animal and human studies, and to right science's tilting hull with neuroscience's long-standing, species-common model of brain, mind, and consciousness. In the process, and perhaps surprisingly for all, neuroscience has emerged as a champion of animal rights with its declaration that animals have the neuroanatomical wherewithal that enables our own species to think and feel.

Along this intellectual journey will be two intertwined themes: predator natural history and neuropsychology. Each chapter focuses on a particular psychological concept and its neurobiological foundations to explore the minds and experiences of a given predator species or constellation of species with an emphasis on those who hail from North America. The specific animals discussed here do not constitute a comprehensive list; instead they are examples that illustrate how discoveries about the human psyche might be used to deepen our understanding of carnivores and help restore their lives and self-determination. As we make our way through layers of misperceptions, it will become obvious that the most salient feature connecting these widely varying species is that they are all targets of human persecution.

In the way I present the stories and science, I also hope to provide a new pedagogical model, one that reinstates subjective experience as a legitimate source of data in partnership with science's accumulated wisdom. The two are, after all, intrinsically compatible since objectivity is merely a collectively agreed-on subjectivity.

Every chapter is grounded in biology, but much of the validating data is provided by those whose schooling draws less from books and formulae than from "watching and wondering." Neuroscience and natural history form the backbone of this work, but it is the free-diver, armed with only a single breath, swimming inches from the lightning-fast jaws of a white shark; the woodsman, weaponless, carrying only a poplar staff, walking beside a mother grizzly and cubs; a biologist, welcoming a rattlesnake in joyful, auric embrace; and a man in shorts and hat nursing the wounds of a fourteen-foot American crocodile, who have brought the greatest insights. These trans-species pioneers are courageous not because they interact with powerful animals in the wild, but because they have stayed strong in weathering the judgment of fellow humans, which, even though verbal or symbolic, can prove just as deadly.

Although not trained as psychologists, these individuals intuitively function in this capacity. Their fresh, unbridled methods showcase a pedagogy akin to what Nikolaas Tinbergen called *scientia amabilis,* a philosophy and practice of thoughtful, loving observation that have since been replaced by the grinding machinery of science's post–World War II industrialization and ask-no-questions objectification. In lieu of modern science's "mark of the plural" that depersonalizes its subjects, Linnaeus, Tinbergen, and their ilk focused on the individual frog, raven, and stickleback to learn about the species, but did so without drowning them in what essayist Albert Memmi described as "anonymous collectivity," sweeping generalities of "They are this." "They are all the same."[1]

This exploration into the lives of carnivores offers two messages. First, if conservation only preserves wildlife numbers and bodies without tending to wildlife minds and societies, then it will fail. If they survive, elephants, grizzly bears, and orcas will exist as mere shells unless their souls are nurtured. The second message hits closer to home. For by vanquishing myths that mask the true identity of carnivores, neuroscience reveals a disconcerting truth: it is not the bear, shark, or crocodile who possesses the villainous qualities of which they are accused, but our own species. The terror does not lie without, but within.

Both of these revelations, taken together, offer us a chance to re-envision wildlife and our identity as a species, and to craft a thriving, more humane culture that no longer demands killing as proof of life. In its place, we are free to return to a gentler, ethical way of being in the world, one in which our carnivore kin are given the kind regard and respect that they have shown us all along.

Introduction
A Tension of Opposites

When studying an individual . . . you enter his or her world as fully as possible. You try to break down your personal barriers of culture, upbringing, education, and open your pores to his or her reality. You try to see and hear on a sub-conscious level.

—Harley Shaw

Harley Shaw's statement is an unremarkable truism among psychologists: it expresses the basic ethical and practical philosophy of all mental health professions. To understand someone, especially someone who hails from a very different background, it is necessary to withhold judgment, to stand in his or her shoes and see the world through his or her eyes, to empathize—indeed, to almost become the other individual.[1] Yet when the full context

and author are revealed, the statement *is* remarkable, because the "individual" is actually a puma, and the author is an Arizona wildlife biologist who has studied the elusive species of American lion for over thirty-five years.[2] The cross-species resonance with human psychological practice is even more noteworthy because Shaw, like most biologists, is not psychologically trained.

For some, focusing on subjective states of animal minds treads too close to anthropomorphism—the attribution of human qualities to another species—and so violates scientific objectivity. Psychology and animal behavior were developed as separate fields for a reason. Only humans were presumed to possess minds, hence a need for psychology. By contrast, animals, who were assumed to run on programmed instincts alone, were considered adequately served by ethology.

But over time, a steady accumulation of science has shown otherwise, and the lines between species have become increasingly blurred. Ravens use tools, horses read human facial expressions, elephants mourn, octopi enjoy unique personalities and moods, and so on throughout the taxonomic tree.[3] All of this gives credence to what Charles Darwin maintained a century and a half ago: differences between human and animal mental lives are more of degree, not kind.[4]

Nonetheless, a reticence toward considering nonhuman species as having similar mental lives has lingered. While human and animal studies have drawn closer, their convergence has remained asymptotic. That is, until neuroscience stepped in with a stunning declaration: all vertebrates (and, it appears, invertebrates) possess common neural substrates that govern brain, mind, and consciousness.[5] In other words, human and animal brains exist and function like variations on a common theme.[6] This discovery showed that past models of distinct human and animal mentation were artificial, as one of psychology's progenitors, Sigmund Freud, himself observed: "In the course of the development of civilization man acquired a dominating position over his fellow-creatures in the animal kingdom. Not content with this supremacy, however, he began to place a gulf between his nature and theirs. He denied the possession of reason to them, and to himself he attributed an immortal soul."[7]

Although neuroscientists still have considerable territory to plumb, the breadth and depth of their studies have provided a much-needed foundation for exploring the less material realms of thoughts and feelings. Sufficient understanding of core brain functions allows us to comfortably make cross-species inferences, even regarding that most human of all qualities, a sense of self. Brain researchers declare that there is "a common form of neural coding across species in subcortical and cortical midline regions in terms of the neural processing of self-relatedness" shared by a wide range of animals—from introvert gorillas, murmurating birds, gliding, graceful sea horses, bats, cats, rats, and scintillating fish—all of which has led to the founding of a trans-species psychology.[8]

Nowadays, scientists are generally much more relaxed talking about their animal subjects. What was once criticized is acceptable. Scorned for his personal touch with African lions, George Adamson is now praised for providing "truths about the species that cannot be found in a biologist's notebook." In fact, researchers from diverse fields have called for scientists to shed what primatologist Frans deWaal refers to as "anthropodenial," a "blindness to human-like characteristics of other animals and animal-like characteristics in ourselves." In its place, herpetologist Gordon Burghardt and others suggest that science employ "critical anthropomorphism," whereby researchers of the mind and brain recognize cross-species similarities, but are careful to avoid projection, that is, imposing their experiences onto their subjects.[9]

The culturally imbued view of humans as the only species with a mind and in possession of unquestionable superiority, however, has created a defensive wall that has proven nearly impenetrable even to neuroscience's data-laden findings of profound psychological similarities among species.[10] For many steeped in the past's conventions of *scala naturae* and the triune brain, which situate nonhuman animals below humans, not only are the concepts of neuropsychology unfamiliar, but their political and philosophical implications are challenging as well.

For traditionalists, Shaw's description of how we might break down our "personal barriers" and become "open" to an animal's reality threatens to muddy the waters of objectivity, something for which Jane Goodall was severely chastised when she named, instead of numbered, her chimpanzees. Any such observations were summarily shrugged off as "anecdotes," unreliable musings without any substantive scientific basis. This negative response to claims of animal sentience has been almost automatic.

Because ethology has been distinguished from psychology, its subjects, animals, have been defined to lack what psychology's subjects, humans, have. Conventionally, the field of animal behavior did not include theory concerning animal mental and emotional states or the common equitable evolutionary substratum that would (as neuroscience does) account for species-mutual expression. Even when cognition has been added, as in the case of Donald Griffin's critical founding of "cognitive ethology" by retaining the root "ethology," two distinct models and fields persist—one for humans and the other for animals. As a result, and by definition, observations of animals grieving, reflecting, judging, dreaming, and all the other subjective states that our species experiences have functioned largely as "add ons" to the basic model of behavior, lacking any fortifying theory that would predict (or contradict) empirical findings. While animals might *seem* to have these human-like capacities, the absence of a concurring conceptual foundation eroded confidence in animal sentience research.

But finally, when neuroscience broke the impasse, nonhuman-human symmetry was acknowledged and a new egalitarian trans-species paradigm emerged.[11] As startling as it may sound to a biologist that a rattlesnake indulges with conspecifics in a Wagnerian-like drama or an orca breaks the likes of an Ephebic Oath, these assertions are entirely consistent with the very tangible science of neural pathways and processes.[12] By grounding animal minds in the substrate of the brain, neuroscientists offer a rigorous means by which field observations reporting animal emotions and mental states may be evaluated. Observations that were scoffed at or dismissed as mere "anecdotes" are, if they are consistent with what neuropsychology predicts, legitimate data.

Since the species-common model asserts that crocodiles, rattlesnakes, bears, white sharks, and pumas have the capacity to feel grief, love, loss, joy, melancholy, and the entire rainbow of mental states that we experience, the question shifts from wondering if a given individual animal has these capacities to considering how, when, and why he or she expresses them. Furthermore, because behavior is only one of the psyche's many expressions and only one of psychology's many measures, ethology is handily married into the disciplinary family of the mind, neuropsychology, so that three disparate fields merge into a single, unified approach to the study of all animals.

Nonetheless, the shift from disciplinary separatism to unification has met resistance, even by those assiduously seeking to liberate the species bias under which nonhuman animals labor. The conceptual and disciplinary apartheid that maintains animal behavior and psychology as independent fields not only contradicts the trans-species science articulated by neurosciences, indeed the entire corpus of science; it also bolsters the perception that nonhuman animals are subordinate and unequal to our species. This has implications beyond academia for law and policy that pertain to nonhuman wellbeing and protection. By not approaching nonhuman animals from the species-common model of brain and mind, ethology loses the powerful inferential leverage that neuroscience offers. And by not embedding observations about animal psychological states and expressions in neuroscience's trans-species framework, the insights of Shaw, Russell, Adamson, Goodall, and others continue to be perceived as *anecdotal* when, in fact, they provide incisive breakthroughs for understanding the inner lives of our animal relatives.

Other kinds of species prejudice persist. Not all animals have been regarded dispassionately. Science hosts on television nature programs may speak empathetically about elephant and emu emotions, but predators generally garner far less sympathy. There is always a shadow side to the story. Tears well upon hearing a tale about a lioness and her bumbling cubs or a fox grieving when his mate is killed, but inevitably these moments are countered by footage depicting what is considered to be much more typical carnivore behavior—a cheetah taking down a baby gazelle or a polar bear tearing apart the body of a lifeless seal. No matter how emblematic the lion, fascinating the wolf, and awe-inspiring the grizzly, an uneasy tension remains. As a consequence, these species occupy a kind of liminal space, variously admired and despised, while both protected and persecuted.[13]

What accounts for what wildlife biologist John Shivik calls the "predator paradox"?[14] People offer various reasons for their ambivalence, but the majority have never encountered a live grizzly, wolf, or white shark. So who *are* these larger-than-life beings that preoccupy human imagination, myth, and mind? A first clue comes from the term "predator."

The classification criteria of predator are a bit tangled. In its most general sense, a predator is defined as an animal who naturally preys on others. Strictly speaking, this would include not only the wolf who eats elk, but the elk who consumes grass, since herbivory and mycophagy (eating fungus) are types of predation. But "predator" is most often associated with those who possess teeth and claws and belong to the *Carnivora* order. Even though carnivorans (members of that order) are generally flesh-eating, the order also includes omnivorous grizzlies and herbivorous giant pandas. Blue whales who use filter-feeding to consume krill and robins who use beaks to gobble up earthworms are also predatory and carnivorous, but they are not carnivorans. Technically conforming to the definition of predator, blue and most other whales are generally regarded as prey, because they are hunted by humans and orcas. So it is not clear which criteria qualify one animal or another as predator.

Obviously, some other factor is responsible. We find clues in the cultural myths that thread through literature, art, and film. Think of the great white shark on steroids, the gaping-mouthed hulking grizzly, or the mountain-sized dino-lizard who tantalizes us with fingernail-biting thrills. *Jaws, In the Maze,* and *Jurassic Park* are blockbuster films because their stars are scary. *Peter Pan*'s lurking Tick-Tock crocodile, *Harry Potter*'s slithering snake, and *The Hobbit*'s snapping, snarling wargs (Old Norse for "wolves") shake a cautionary finger at careless children. And, we are warned, even though lions and tigers look cuddly and huggable, they are not trustworthy. They will suddenly, for no reason, lash out and kill even those they have known for years. Consider the white tiger from the Siegfried and Roy Las Vegas show who almost killed one of the performing duo with a bite on the neck. Sooner or later, we are told, predators revert to their true nature as killing machines. Are these assertions true, or is there something else to these stories? When examined more closely, reasons used to justify human suspicion and unease don't line up with the facts. Something much more complex lies at the bottom of humanity's love-hate obsession with carnivores.

Reason 1: Carnivores stalk and eat other animals. This is true. That carnivores hunt for and eat other animals is certainly a factor in the attraction versus repulsion equation. Wolves, white sharks, and pumas are what biologists call obligate carnivores or hypercarnivores, that is, they are evolutionarily adapted to require flesh to survive and thrive. But, as we saw, not every species on humanity's most-wanted list has such a restrictive diet. Tiger sharks are quite happy to scavenge on spoils left by another; the majority of grizzly bear diets include grains, legumes, and berries; and frugivory is "widespread" among the *Crocodylia.*[15] So it is not just carnivory that makes humans wary of these species.

Reason 2: Carnivores are cruel. No matter how it's framed, the charm of *Born Free*'s Elsa the lion fades a little when viewers watch her running down and chomping on a baby antelope. Yet when man's best friend, the dog, chases a neighborhood squirrel, or during the heat of an organized hunt tears apart a still-living fox, he is exonerated, unjudged and unpunished, with his murderous streak shrugged off as wolfish ancestry. For that matter, few notice the differences between humanity's ritualistic mass killing of turkeys, who are doomed to suffer from birth to death in the confines of cages, and the parsimonious killing of a white-tailed deer by a puma's stiletto teeth.[16] An ethical double standard leaves carnivores consistently on the losing side.

Reason 3: Carnivores are mindless killing machines. Here, taxonomic prejudice comes into play. Some might say that Cecil the African lion was perhaps at times cruel—but unfeeling, no. Nor would many cast other than a fond eye at bottlenose dolphins and meerkats, who munch on squid, crabs, fish, and birds and, in the case of friendly Flippers, sometimes engage in not-so-nice habits such as infanticide. Nonetheless, they remain darlings of the media. Meanwhile their fish counterparts, sharks, are denied any such largesse. Western Australia's Premier Colin Barnett drew a distinct line in the taxonomic sand with regard to the state's new "kill on sight" shark policy when he told his constituency, "This is, after all, a fish—let's keep it in perspective."[17]

While most fish are not considered predators in the same way that coyotes, wolves, and pumas are, details of the fish sentience debate provide key insights into the tangle of predator issues. For instance, there are powerful reasons for keeping fish and other animals in dumb and dumber roles. If, as the weight of evidence shows, fish feel pain, German law qualifies them for protection that "levies severe fines or jail time for anyone who willfully, without undue cause, inflicts pain or suffering of a sentient being."[18] Fish sentience has "far-reaching consequences for millions of anglers, fishers, aquarists, fish farmers and fish scientists" who depend on being able to use finned species in ways that involve inflicting pain.[19]

For many years, the topic of animal sentience had difficulty gaining scientific traction. Discussions and debates on whether fish, bats, cats, and other animals could feel and think frequently ended with a frustrated "it's a matter of opinion." Now with neuroscience in the mix, the frontlines of the sentience battles have extended to the trenches of neuroanatomy. Those claiming that fish don't suffer the way we do maintain that the "conscious experience of pain most likely requires highly-developed and regionally specialized forebrain neocortex and associated limbic cortex which fish do not have." Moreover, to "the extent that human brains and fish brains differ, particularly the great differences between human neocortex and limbic mesocortex, vs. pallial structure in fishes, the properties of putative consciousness in humans and fishes would differ as well." But such protestations drown in an ocean of data showing otherwise: "the absence of a neocortex does not appear to preclude an organism from experiencing affective states," and neuroarchitectural differences among species do not detract from functional comparability. Fish (and bird) brain structures correspond to those of mammals.

Although singularities in brain features such as Mauthner cells in the spinal cords of some fish exist, they are rare and the genes responsible for determining main brain structures have changed little since vertebrates diverged from arthropods more than 500 million years ago.[20] Regions of the brain are conserved across vertebrate species—mammals, birds, reptiles, amphibians, and fish—and all "have the neuroanatomical, neurochemical, and neurophysiological substrates of conscious states along with the capacity to exhibit intentional behaviors." Even invertebrates—octopi and insects—are included on the list. Scientists have discovered that bees and friends have consciousness, are self-aware, and have an ego and that these subjective states and experiences nucleated in the Cambrian. The conclusion? "The weight of evidence indicates that humans are not unique in possessing the neurological substrates that generate consciousness."[21] But scientific arguments for cross-species parity don't stop here.

Continuing with the example of our finned friends and the topic of sentience, behavioral researchers have reified what neuroscientists had predicted: fish exhibit a diversity of high-functioning behaviors that correspond to and reflect neural substrates. Fish not only outcompete capuchin monkeys, chimpanzees, orangutans, and human children in foraging tests; they also cooperate, communicate with each other, and exhibit empathy and reflexive self-concern.[22] The strength of these now-indisputable piscine neuropsychological data has convinced primatologists that "primate chauvinism may now be poised to decline, thanks in large part to . . . fish work."[23] After years of being told that the fishing industry doesn't want "tunas with good taste," only "tunas that taste good," Charlie the Tuna has been vindicated.[24] Predators, whether fish, fowl, or beast, may be killers, but they are mindful.

Reason 4: Carnivores are costly. Carnivore critics argue that livestock depredation—wolves eating sheep, foxes eating chickens, pumas eating llamas, and so forth—creates an unacceptable economic hardship. But when statistics on the numbers of cows, chickens, and other domesticated animals who become food for these predators are examined, it becomes clear that these numbers are vanishingly small and insignificant, particularly when viewed with the understanding that farm animal industries are heavily subsidized by the government.[25] For example, when a study on cattle depredation in Brazil calculated "predation loss rates," it turned out that "jaguars were responsible for 0.83 percent of [the] loss in two ranches in the northern Pantanal. The big cats took 0.3 percent of cattle from another ranch in the southern Pantanal, 1.26 percent in southern Amazonia, and 1.2 percent [in the] study area." Local ranchers insisted that predators had accounted for at least 11 percent of their total losses. Researchers, however, discovered that "it was not the actual damage that was driving the cycle of persecution, but a blighted perception. It was the threat of damage."[26] Reasons for anti-predator bias are "deep rooted and value laden and are connected to individual lifestyles and views of the place of humans in nature."[27] As a result, programs created to compensate for predation such as wolf-related losses of livestock and hunting dogs have often had "little effect on tolerance for wolves suggesting that attitudes toward wolves are not purely a matter of economics."[28]

Reason 5: Carnivores deplete the supply of available game for human hunters. Once again, perceptions in this area fail to match reality. Historically, North American herbivores thrived while living with abundant numbers of carnivores with whom they had co-evolved. Early settler testimonies extol a richesse of birds, deer, elk, and their predators. Prior to colonization, the lands and seas occupied by tribal peoples looked very different than what colonial occupation rendered. European settlers and their descendants, not animal carnivores, have gutted nature's stocks in North America, as on other continents. Daphne Sheldrick, founder of The David Sheldrick Wildlife Trust, Nairobi, Kenya, recounts the bounty of wildlife that her relatives encountered at the onset of Africa's colonization: "My great-grandparents were amongst the fortunate few to have been in time to witness this wonder . . . Then came two world wars, and in the great herds lay an endless larder from which cheap protein for prisoners and troops could be drawn. Game was looked on as vermin in the European settled areas; a threat to a farmer's stock by harbouring disease, and a threat to his land by competing for grazing . . . Life was cheap, the stocks of game seemed endless there for the taking; a source for free meat and free sport for anyone who could hold a gun." Within a very short while, the "great herds began to dwindle, eroded by the impact of civilization, and with each year that passed, the numbers grew fewer, until people suddenly wondered in astonishment where all the animals had gone."[29]

Nonetheless, despite what history documents and that predator "control efforts tend to be ineffective" and costly, wild carnivores are blamed for herbivore decimation.[30] Legislators in the Pacific Northwest know that it is fishermen and hydroelectric dams, not sea lions, cormorants, and orcas, who have pushed salmon and steelhead to the top of the endangered species list. Yet they are bent on passing a bill to kill sea lions who, as they have for thousands of years, must eat fish to survive. Similarly, the U.S. Army Corps of Engineers has initiated a campaign to kill thousands of double-crested cormorants and their eggs because the species feeds on salmon.[31]

Reason 6: Carnivores serve no purpose. From the perspective of modern utilitarian culture, carnivores present a problem. They don't make "good eatin'," and they can't be domesticated for any useful purpose. For a long time, their skins were valued, but that only carried them so far. While the flashy tiger and lion still pull in lucre for zoos and circuses, most live carnivores don't provide humans the same feverish attraction as killing does.

In contrast, hunting wildlife is big business. In 2011, U.S. hunters spent $34 billion on their pursuits. This does not take into account ancillary economics: that is, diverse business interests that include motels, restaurants, weapon sales, and clothes. These interests reportedly generated $11.8 billion in taxes over the year, funds that are used to support government wildlife agencies.[32] While an estimated $200 million annual revenue from African lion trophy hunting supports local conservation, studies reveal that "only about three per cent of these fees actually reaches the local communities. Most of the money is siphoned off by the hunting industry and government officials." Similar statistical patterns are found in the United States. A study investigating the assertion that hunting pays for wildlife conservation concluded that approximately 95 percent of federal funding, 88 percent of nonprofit funding, and 94 percent of total funding for wildlife conservation and management come from the nonhunting public.[33]

Reason 7: Carnivores are dangerous to humans. Every year, bear, puma, raccoon, coyote, skunk, shark, and other "nuisance" wildlife populations are obliterated based on the claim that doing so is necessary to preempt human harm and fatalities. If the precautionary "kill or be killed" adage is taken seriously, however, then it is more logical to place a bounty on bathtubs. Each year there are, on average, 20,000 deaths, 7 million disabling injuries, and 20 million hospital visits resulting from accidents within the confines of American homes.[34] Approximately 4.5 million people are bitten by dogs each year. Between 1982 and 2014, there were 4,870 "fatal and disfiguring" attacks by dogs, resulting in 579 deaths, and annually, deer, whose historical migratory paths are now severed by roads, "fling themselves in front of 247,000 motor vehicles and injure or kill 26,647 drivers and occupants." Furthermore, human-on-human deaths eclipse all of these statistics. The U.S. federal law enforcement agency reported that an estimated 1,165,383 violent crimes (murder and non-negligent homicides, rapes, robberies, and aggravated assaults) occurred in 2014.[35] This figure does not even include fatalities from wars.

In contrast, there were twenty confirmed puma-related human deaths over a 120-year period, which amounts to 0.17 deaths per year. Three fatal shark attacks occurred worldwide in 2014 compared with 1.24 million vehicular deaths in that same year. No more than five people annually succumb to the fangs of a rattlesnake. Even in the nineteenth and early twentieth centuries when North America could still claim extensive, relatively untouched tracts of land occupied by dense populations of wildlife, there were, as John Muir observed, "many snakes in the cañons and lower forests, but they are mostly handsome and harmless. Of all the tourists and travelers who have visited Yosemite and the adjacent mountains, not one has been bitten by a snake of any sort."[36] Coyotes rarely bite, let alone kill and eat humans, yet they remain atop the most-wanted pest list. And compared with other deadly threats, the number of African lion attacks on humans is vanishingly small. In Tanzania, "193 to 1,499 people per year die of rabies-infected dog bites, 600 from snake bites, 1,900 from falls, 4,700 from drowning, 6,000 from asthma, 13,000 from road accidents, 14,000 from violence/homicide, 21,000 from malaria, 23,000 from diabetes, 35,000 from diarrhea, and 122,000 from HIV/Aids/tuberculosis."[37] The disparity between fatalities caused by wildlife and those caused by other agents is even more astounding given that bears, big cats, and even marine carnivores must survive in a fraction of their historical habitat, with a hugely reduced resource base that is often in close proximity to human communities.

Others point to an eighth reason for predator panic, one that is more subtle and unconscious. The "underlying emotion that sustains the partisanship [against carnivores] is rarely expressed or acknowledged, but it is the primary motivator for advocates and eradicators alike . . . It is the whispered language between predator and prey. It is fear." After surviving three death rolls in the arms and jaws of a saltwater crocodile, Val Plumwood became well acquainted with this feeling. Hers was a terror born not only from the specter of death, but also from the outrage, the profanity, that the "creature was breaking the rules." The reptile who attacked and nearly killed Plumwood was "totally mistaken, utterly wrong to think I could be reduced to food. As a human being, I was so much more than food."[38] This sense of shock is credited to an "implicit corollary to the Fifth Commandment, one that is incumbent upon the victim rather than upon the aggressor: Thou shall not be killed by being consumed."[39] By upending the trophic chain, carnivores challenge humanity's self-proclaimed, privileged authority.

In summary, then, those species who are variously referred to as predators, carnivores, or pests are grouped together not for a single reason, but rather due to an amalgamation of motives that have more to do with human attitudes and psychology than facts or biology. This prejudice is not whimsical, nor is it easily assuaged. Resentment toward carnivores runs deep.[40]

When Europeans claimed new worlds, they "carried within them generations of myth, fear, and violent reaction to predators."[41] As far back as the 1500s, Jesuit priests offered bounties for killing pumas—one bull for every catamount, as the big cats were called.[42] Theodore Roosevelt cemented this sentiment when he damned the wolf as a "beast of waste and desolation" and the puma as "the big horse-killing cat, the destroyer of deer, the lord of stealthy murder, facing his doom with a heart both craven and cruel."[43]

The spitting rhetoric eventually solidified into ironclad American law as wildlife species were increasingly pulled under government authority. In 1885, the Division of Economic Ornithology and Mammalogy was established in the Department of Agriculture, which was, in 1905, expanded and renamed the U.S. Bureau of Biological Survey.[44] In 1915, the U.S. Congress authorized the Division of Biological Survey to systematically kill all wolves, coyotes, and other "predators"—and so they did, with efficiency.[45] By the early 1900s, virtually all pumas east of the Mississippi had been killed.[46] By the 1930s, wolves had been vanquished from most western American states, and grizzly bears had been driven from 98 percent of their coterminous habitat.[47]

The pressure on carnivore populations has only increased. In addition to the loss of habitat to agriculture and development demanded by a skyrocketing human population, trophy hunters and government-sanctioned culls continue their escalating assault. Since the 1960s, African lion populations are estimated to have fallen from over 250,000 to as low as 15,000, with scientists claiming that the possibility of the species' extinction is very real and not so far away. While the debate on the number, or even existence, of tiger sub-species is ongoing, there is consensus that tiger genetic diversity has been greatly diminished. Tigers in the wild number fewer than 3,900. Eight shark species declined by more than 50 percent with white and hammerhead shark populations dropping by 80 and 90 percent, respectively, over the past fifteen years.[48]

Even carnivorous reptiles are struggling. The loss of Chinese alligator habitat is nearly total, and the species is expected to become extinct in a few years.[49] Godzilla's real-life cousin, the Komodo dragon, has been able to hold its own, but conservationists are now scrambling to stop mounting threats to the lizard's survival.

Less charismatic predators such as raccoons and skunks may have better maintained their numbers, but they are killed en masse annually simply because they trespass areas of human occupation. In 2012, it was reported that since 2000, employees of the federal Department of Agriculture Wildlife Services "have killed nearly a million coyotes, mostly in the West. They have destroyed millions of birds, from nonnative starlings to migratory shorebirds, along with a colorful menagerie of more than 300 other species, including black bears, beavers, porcupines, river otters, mountain lions and wolves."[50] Carter Niemeyer, a former Wildlife Services district manager who worked for the agency for twenty-six years, believes that "much of the bloodletting is excessive, scientifically unsound and a waste of tax dollars. 'If you read the brochures, go on their website, they play down the lethal control, which they are heavily involved in, and show you this benign side . . . It's smoke and mirrors. It's a killing business.'"[51] Predator persecution has become so extreme that it has earned the eerie title "carnivore cleansing."[52]

Beyond negative economic effects, misplaced culpability has had devastating ecological consequences. Scientists protest that "eliminating large carnivores is one of the most significant anthropogenic impacts on nature," and is a crisis that is ecological as well as ethical. In the attempt to solve a problem that does not exist, "carnivore management" has unleashed a chain of real problems. Without the regulatory strong arm of large carnivores at higher trophic levels, the cascading dynamics governing the resilience of "meso" carnivores, plants, and other animal communities will spin out of control. A recent headlining analysis on key carnivores showed that such losses "can be expected to influence numerous . . . ecological processes, including disease dynamics, wildfire, and carbon sequestration . . . [with] wide ramifications through highly interconnected food-web networks within their associated ecosystems."[53] The nosedive to extinction of "more than three-quarters of the 31 species of large land predators" has caused tsunami waves down the food chain. Flora and fauna that have co-evolved with carnivores are suffering a suite of maladies and even systemic collapse as a result of human excesses.[54] It is estimated that 30 percent of freshwater reptiles, many of which are not carnivorous but provide sustenance for many large-bodied carnivores, are close to vanishing. This proportion increases to 50 percent when freshwater turtles are considered. Carnivores are in fact essential to the populations on whom they prey.

Meanwhile, amid the proliferation of official finger-wagging warnings, frenzied funding for anti-carnivore initiatives, and fish stories of gaping jaws and lucky near misses, the real animal has been lost. Even when exonerating facts are presented, it seems that adrenaline-laced myths are so seductive, so much more intriguing than the quiet of truth, that few take the time to discover who lies behind the mask of falsehoods. But some have been given pause. Some have tuned out the hiss of lies and dedicated their lives to learning about carnivores from the inside out. These individuals have made friends, not foes, with the toothed and clawed. What they have discovered is that bears, pumas, rattlesnakes, crocodiles, and sharks are not apex predators but rather "apex models of civility and grace."[55]

Until recently, stories of people hugging bears and snuggling snakes were dismissed as sentimental indulgences or the doings of hapless idiots. But the myth that carnivores are mindless killers has worn thin and repeatedly fails the test of facts. Kevin van Tigham, former superintendent of Banff National Park, Canada, reflects on his more than forty years of experience: "In our dealings with bears we too often focus on our fears and fantasies . . . The most dangerous thing about a bear is not its claws, teeth or disposition; it's how we react to it."[56]

Increasingly, scientifically trained individuals, such as van Tigham, are challenging conventional wisdom. Although their views may conflict with what they have been taught, new generations of biologists are finding surprising camaraderie in the fields of psychology and neurosciences, which recognize animal sensibilities comparable to our own. Furthermore, Gordon Burghardt, who has studied a wide range of species from snakes to bears, is one of many now who argue for the importance of subjective experience in study. Not only is this open approach necessary to accurately and fully grasp who the subject is, but "the separation of both experimenter and emotion. . . in scientific studies. . . is just not possible."[57] Scientists have also realized that studying animals from the outside is not enough. As a series of studies on bears conducted in 2015 demonstrates, consideration of animal minds is essential if the goals of conservation are to be served. Numbers and behavior do not tell the whole story.[58]

Recreationalists and researchers are using unmanned aerial vehicles (UAVs, or "drones") in wilderness areas to spy in on wildlife. To see whether these airborne machines were bothering black bears, researchers measured the animals' physiological and ethological responses. While Smokey the Bear's real-life counterparts exhibited "infrequent behavioral changes," the scientists "observed consistently strong physiological responses" in the bears and "all bears, including an individual denned for hibernation, responded to UAV flights with elevated heart rates, rising as much as 123 beats per minute above the pre-flight baseline."[59] Similar to studies on humans, ethological analyses had yielded a false negative.[60] Anxiety and fear—psychological states of mind—had caused a pronounced physiological response in the bears.

The inclusion of psychology in conservation has shown its importance elsewhere. Young wildlife are not acting like they have in the past. Notoriously covert, young male pumas are now being sighted in cities and suburbs.

While some may attribute these behavioral changes to the keen adaptive abilities of the species, attendant symptoms—such as unprecedented puma inbreeding and cases of intraspecies killing, which have accompanied habitat destruction and high mortality rates from hunting and vehicles—point to a more expanded explanation.

The oceans are having their own at-risk youth problems. Teenage male dolphins are running down and killing porpoises and conspecific infants street-gang style. Typically, although evolutionary strategies are called on to explain infanticide, biologists have been hard-pressed to explain the dolphins' motives. But to a neuropsychologist the most obvious and likely cause in the puma and dolphin cases is psychosocial trauma. Changes on the outside that have been sustained by wildlife for generations—mass killing, loss of habitat and adequate prey, perennial lethal pursuit—eventually leave their traces on the inside. Even the fiercest creature has limits.

The predator story is not an easy one to hear, even for carnivores' advocates. It is challenging to find love and light in the deadpan eyes of a circling shark or a lynx playing with her still-living, terrified rabbit prey. But as Harley Shaw points out, it is neither scientific nor fair to judge another without first stepping into their shoes. Life in the armchair of modernity where a flip of the switch turns on the heat and warms up a meal from an ever-handy fridge is a far cry from amenity-free outdoor living. And when it comes to judgment it is modern humanity, not carnivores, that warrants ethical scrutiny and reinvention.[61] In the rush to blame someone else, attention has been diverted from the most dangerous predator, humans.[62]

Neuropsychology provides a new key to unlock our understanding of other species' experience in our shared world. This interdisciplinary, species-spanning model of brain, mind, and behavior compels us to step beyond dualism's demand for black-and-white, either-or classifications and learn how to accept other species' differences and similarities—that is, to embrace how we are the same as other animals and, at the same time, different. In so doing, we may open up an opportunity to find out who our species can become. It is an alternative reality that harks back to the ancient past when the coyote's howl, snake's rattle, and grizzly growl inspired awe instead of fear and loathing. The grizzly countenance peering from the pines, diamond-shaped visage nested in his coil, and circumspect feline mien are not the faces of strangers—they are kin. The toothed and clawed beckon us to heed Harley Shaw's call to look beyond the hunt to the common heart we all share as inhabitants of the Earth.

I

White Sharks: Personalities

In 1942, on the eve of leaving for an OSS assignment in the Balkans, my father gave my mother a gold charm for her bracelet: a tiny rectangular placard engraved with "If you can read this, then you're too d—n close." The charm's caution came to mind when I was told that the iris of a great white shark is sapphire blue. To ascertain shark eye color, you have to be *pretty damn close.*

It's true. The real-life relatives of *Jaws* have the same eye color as Bette Davis, Paul Newman, Jonathan Rhys Meyers, and "Old Blue Eyes," Frank Sinatra. Otherwise, sharks give the appearance of what has become the classic description of a dangerous psychopath, whether fish or human: a blank,

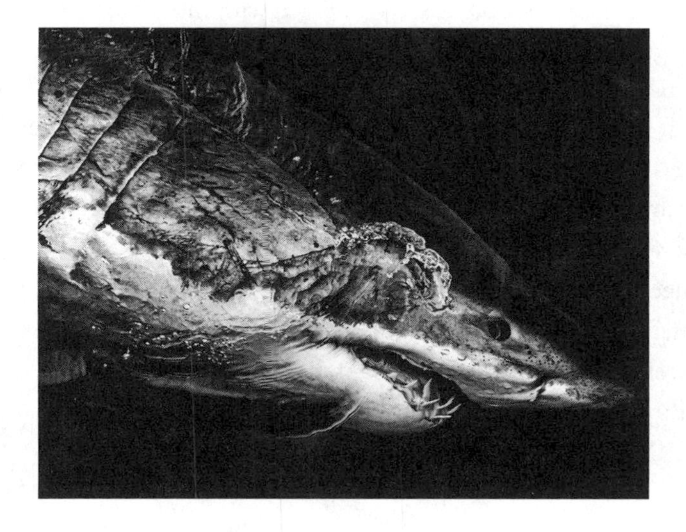

deadpan stare. To their credit, much of this unsettling look comes from the very unmammalian feature of having lidless eyes. Laminid sharks, such as the white and mako, roll back their eyeballs instead of shutting a protective lid. But for seasoned diver-photographer Fred Buyle, eyelid or no, connecting with a shark's windows to the soul is an essential element for survival. The ability to swim as close as he does to that startling blue eye, not once, but thousands of times, requires understanding the psychology of the enigmatic white shark down to every nuanced detail.

Water has been Buyle's world since he was four years old. At thirteen, he had his coming-of-age encounter with a reef shark while diving in the seas of Indonesia. This experience launched him on a career in free-diving competition, and in 1995 he set his first world record. He achieved three more world records between 1997 and 2000, which included, in 1999, becoming the eighth person in the world to pass the "mythical barrier": use a single lungful of air to plunge down more than one hundred meters and return to the surface.

But Buyle's years of honing the art of free-diving have been motivated by something more than competition and achievement. They were a practical means to a more sublime end—being able to meet with denizens of the deep on their own terms. As he puts it: "Each dive is a mission and each is a journey to a different planet."[1] Of the over four hundred (some estimate five hundred) shark species, Buyle has free-dived with more than a score.[2] Today, holding his breath for several minutes underwater is second nature, but he insists that with practice, anyone can do it. "The sport is made out to be so special. It is in the sense that it is a beautiful and extraordinary experience. But everyone can learn. That's what I teach. It's natural. The water is where we all came from." Buyle is an invaluable resource for researchers seeking to probe what the ocean blue covetously hides. He works with a wide range of engineers, scientists, cinematographers, and other free-divers dedicated to protecting perilously endangered marine life. The innovative Darewin Project, which seeks to understand the complex language of sperm whales, is only one of Buyle's many conservation activities.[3]

(Overleaf) The blue-eyed White Shark. Photo credit: Edgar Eduardo Becerril-García.

With the aid of only his practiced breath, a talented free-diver provides a quiet underwater eye that someone in noisy and bubbly scuba gear cannot. Bubbles are an important medium for communication among many marine species. Humpback whales surround herring schools with a self-made bubble net to scoop up the little fish by the mouthful. Right whales emit bubbles for another reason: clouds pouring from the whale's blowhole are part of a male's repertoire for courting the female of the species. A diver's scuba bubbles thus may be mistaken for an untoward communiqué. By sidestepping this issue, free-divers are able to gain the trust of marine dwellers.

The free-diving profession has a long history. Records date back to 5,400 B.C.E. when the Stone Age fishing and diving culture Ertebølle existed. Many societies relied on diving for military espionage, salvaging, and foraging. Free-diving inspired John Steinbeck's novel *The Pearl.* Its popularity as an extreme sport took off starting in the 1960s. Since then, free-diving has grown from a limited circle of experts to a full-fledged sport featuring multiple competition events and organizations governed by the International Association for Development of Apnea (AIDA). The current world record for diving with a single breath is seven hundred feet below the surface.

Among all the planets in the watery universe, there is one that almost everyone is tempted to visit, even if perhaps not in the intimate ways in which the Belgian diver has: the world of the legendary white shark (*Carcharodon carcharias*). Buyle met his first white in 2007 off the shores of South Africa. Along with Guadalupe and Australia's reefs, southeastern Africa is one of three traditional hotspots for viewing white sharks. The moniker "white" refers to the shark's pale underbelly—the rest of the body is smooth slate gray, though at times it appears almost black.

According to Ralph Collier, president of the California-based nonprofit Shark Research Committee and director of the Global Shark Attack File (GSAF), the descriptor "white" is purported to derive from the 1700s. The story goes that when schooner sailors pulled the big fish on his back to shore, the Swedish zoologist cum botanist Carl Linnaeus was confronted with a "girthy, robust, shark with a big white belly and white pectoral fins," which perhaps led him to overlook the shark's dark topside and assume that the shark's entire body was pale. Hence, the species name.[4]

The White Shark. Photo credit: Fred Buyle.

White sharks loom monstrously large in human minds, but they are not the biggest of their class, *Chondrichthyes* (cartilaginous fishes). Their mild-mannered filter-feeding cousins, whale sharks (*Rhincodon typus*), take the size prize by reaching an awesome twelve meters and nearly twenty metric tons. Another relative, the basking shark (*Cetorhinus maximus*), is only slightly shorter and lighter.[5] Nonetheless, the obligate carnivorous white shark enjoys his status sitting atop the trophic fish pile and, for the most part, lives unchallenged save for the ubiquitous *Homo sapiens;* on occasion, another, bolder shark; or the killer whale (*Orcinus orca*).

Size aside, there is hardly any visage more fearsome to the human eye than that framing a white shark's cavernous, razor-teeth-ringed mouth. The reputation of the species is not improved by the fact that these sharks swim with their mouths open, closing them only when they are about to pounce on their prey. Every year, thousands of tourists flock to see the frightening mouth in action. Shark feeding has become a multimillion dollar industry as far and wide as there are sharks. Commercial boats motor to white shark haunts that provide passengers with a real-life *Jaws* experience: the spectacle of gaping mouths snapping and gulping down chunks of dangling fish. For those made of sterner stuff, there are cage dives. Customers are fitted with suits, tanks, and a guide—then, in an aluminum cage, are lowered into the brine where they can cringe and ogle at sharks grabbing fish gingerly proffered through battered bars. Tourists with yet more moxie have the opportunity to dive into Mexican waters sans cage and hold out morsels to bull sharks (*Carcharhinus leucas*), who relieve them of their fleshy offerings.

Staged viewings yield only one-dimensional glimpses into the depths of shark minds. Historically and still today, most contact with sharks takes place at the end of a hook. This skewed view fuels the bad press that bases its ideas about sharks from photos showing lifeless bodies posed on a sheet of blood or strung up by the tail, gallows style, wreathed by smiling dockside conquistadors. And, despite their larger-than-life presence in the media, sharks have not been the subject of a great deal of scientific research. Consequently, almost every day seems to bring startling information about even the most basic biological facts, such as the water temperatures that sharks occupy. Unexpected sightings revealed hammerhead (of the family Sphyrnidae) and silky (*Carcharhinus falciformis*) sharks carving their way through hot, acidic currents above the submarine volcano Kavachi, located in the Solomon Islands. This discovery surprised shark researchers who never suspected that these species could withstand such hostile waters.[6]

The main reason that so little is known about sharks is the nature of where they live. Unlike land-dwelling animals, marine wildlife has had the ocean's dark, deep blue to safeguard their privacy. The vastness of sharks' potential habitat—oceans cover 70 percent of the planet—and the limitations of surface viewing have forced researchers to patch together bits and pieces of information to obtain any kind of coherent shark storyline. Now, however, scientists are beginning to outfox the ocean. Researchers use transmitters inserted into shark dorsal fins and bodies to keep an eye on the fish day and night. Satellite tagging and associated innovations have enabled shark conservation organizations to show the public a different side of white sharks. For example, fans of Mary Lee, a full-sized female white shark who prowls off the northeastern U.S. seaboard, can watch her progress via online GPS as she zigzags her way from Virginia to New Jersey and the New York harbor.[7] She even has her own Facebook page.[8]

When data from a potpourri of tracking devices—archival, pop-up, and satellite—are strung together, a much more detailed picture emerges showing that there is a lot more to a white shark's life than killing. One significant revelation provided by these technologies is that the range of white sharks is more extensive than previously assumed. While they may show fidelity to a given feeding area, whites spend a lot of time on the move. Mary Lee traveled 19,474 miles in the first two-and-a-half years of observation.

White sharks are considered to be picky eaters and the supposed reason is that they need specific kinds of food to efficiently convert themselves from a cold-blooded fish into an endotherm. The orange and pink, disc-shaped opah (*Lampris guttatus*) is the only known fully warm-blooded fish, but white sharks achieve a similar result through an ingenious method of thermal regulation.[9] Along with its Lamnidae cousins, mako longfins (*Isurus paucus*) and shortfins (*Isurus oxyrinchus*), porbeagle (*Lamna nasus*), and salmon sharks (*Lamna ditropis*), white sharks are "regionally endothermic," that is, they are able to raise and maintain their internal temperatures to as high as 45°F (25°C) above the ambient water temperature. White sharks are prime examples of fish heterotherms—species that can toggle between ectothermy and endothermy. Heterothermy in sharks, tuna, and billfish is "part of an integrated high-performance design."[10] Wading birds accomplish a similar feat by redirecting blood inward using *rete mirabile.* These "wonderful nets" of vessels carry cool, oxygenated blood from the gills via tiny arteries, and in reverse, complementary veins bring back warmed deoxygenated blood. Countercurrent flow effectively recycles heat with the transfer from veins to arteries. This physiological mechanism also explains why sharks keep their mouths open: it guarantees a steady stream of incoming water through their gills, which enables them to breathe. There is a second reason for the gaping maw: an open mouth is always ready to attack and kill prey.

There are many advantages to being warm-blooded. A warm interior may provide an extra boost for sharks' embryonic development in otherwise cold surroundings. It also allows white sharks to do what they do best—hunt. Endothermic capacity dramatically increases muscle contraction speed, which supports the torpedo-fast acceleration needed to strike, bring down, chew, and swallow prey efficiently. Short-fin makos are considered to be the fastest of all sharks, reaching speeds of up to eighty kilometers per hour, but the white's size and speed combine to make the species an almost unbeatable competitor. With the help of their powerful tails, the white sharks' energetic faculties propel them up to speeds of fifty kilometers per hour, which enables them to make their famous surging, whale-like breach when they rise from the depths to catch unfortunate prey.

The "men in the gray suits" (also called by surfers "The Landlord") can even catch the speedy tuna, and there are reports that their strong, muscular jaws can bite through tough sea turtle shells.[11] Lucky for the Mock Turtle and kin, white sharks generally do not seem to be interested in turtles. Although Buyle and others have observed white sharks "checking them out," and turtle remains in white shark stomach contents have been documented, it is the tiger shark (*Galeocerdo cuvier*) who is inclined toward the shelled ones. Buyle explains: "Tigers are the big turtle eaters. I've seen what these guys are able to do with their jaws, nothing can resist those teeth and muscles. Plus, they are built like tanks. On the other hand, a white shark can handily slice a huge tuna in half in the blink of an eye. Each species has different teeth and different lifestyles and different psychologies. But one thing they have in common is that they are perfectly designed for what they need to do. And they are very canny, at times hanging out at fishing operations or grabbing a marlin or other fish that has been hooked by a fisherman and snapping it up, in and away."[12]

Endothermic machinery in the shark expands the climatic range in which they can roam. For the most part, white sharks are found in cool temperate waters near shore in the epipelagic zone—the upper part of the water column—but they have been spotted in the tropics and at times even the frigid waters of the northwest Pacific off Alaska and Canada. In the 1980s, a white shark was observed feeding on walrus in the Bering Strait and as far south as Campbell Island, off New Zealand. In the latter case, the discovery

Blue Shark in blue oceans. Photo credit: Fred Buyle.

occurred when a diver, who was part of a team of scientists installing an ocean floor monitor, was attacked.[13]

Sharks use a multiplicity of senses to detect who is nearby and keep track of where they themselves are located.[14] Typically, much like a trout hunting a fly on the surface of a stream, sharks approach their prey from below. When a shark has ascertained that a strike is worthwhile, she will propel up through the water vertically and grab her prey. A vertical approach maximizes stealth and optimizes the shark's excellent color vision.[15] Most information on sharks describes their movement in the horizontal plane, but tracking technologies are making it plain that the fish are adept users of all the space in which they swim.[16] White sharks utilize a sun-tracking predation strategy that has several potential advantages, including enhanced concealment and improved detection of prey.[17]

There are further adaptive advantages to the white shark's physique. Unexpectedly, these non-direct-air-breathing sharks exhibit positive buoyancy because of their very useful livers. Livers serve multiple vital functions including the storage and release of vitamins and metabolic heat. They are thick with low-density oils and, critically, the hydrocarbon squalene ($C_{30}H_{50}$), which is much less dense than seawater. Low-density liver compounds buoy the shark so that a mere flick of the caudal fin propels her forward. Shark livers may constitute more than 35 percent of body weight.[18] Deep-sea sharks such as the bluntnose six-gilled shark (*Hexanchus griseus*) who have been found at depths of five thousand feet or more rely on such large, oil-filled livers to glide vertically up the water column.[19]

Other shark species take advantage of this adaptation. When the Princeton-based nonprofit Shark Research Institute (SRI) began satellite tagging whale sharks in March 1998, researchers were astonished to find that the sharks routinely dived to depths greater than 2500 feet. One day, at the Explorer's Club, Marie Levine, founder and executive director of the Shark Research Institute, spotted a photo taken in the 1920s of classic western author Zane Grey with a whale shark. This prompted Levine to research further. Once, off the coast of Mexico, when Grey harpooned a forty-nine-foot whale shark, the fish "promptly went into a vertical dive, taking 1,640 feet of line before dislodging the harpoon. Had we known that the whale shark could swim at such depths, we might have anticipated the results of our findings in 1998."[20]

This silent gliding may increase a white shark's stealth as she surprises her prey from below, whether that prey is a seal or a surfer astride his board with wave sets on his mind. Attacks on humans are, of course, the grist of all titillating stories that surround white sharks. But when statistics are examined, sharks, including the white shark, are a near-to-nothing threat to humankind. The GSAF shows that from 725 B.C.E. to 2015, there have been a total of 4,302 unprovoked shark attacks with 1,121 fatalities.[21] Given these numbers, it is reasonable to assert that "humans are not on the menu of sharks, rather, sharks bite humans out of curiosity or to defend themselves."[22] Speaking specifically to the habits of "The Landlord," Collier, also the author of *Shark Attacks of the Twentieth Century*, states that over the past century, there were 108 authenticated, unprovoked white shark attacks along the Pacific Coast of the United States, and of those, only eight were fatal.[23] "When you consider the number of people in the water during that hundred year period," he notes, "you realize that deadly strikes are very rare."[24] Because there is often uncertainty about which species was responsible, estimates for white shark attacks may also include some identification errors. Nonetheless, the numbers are a mere fraction of any cited total.

As with most aspects of shark natural history and psychology, there are diverse ideas about whom, why, and when a white shark will attack. The most common myth is that white sharks, with their poor vision, attack divers and surfers in wetsuits because they mistake them for seals and sea lions. R. (Rick) Aiden Martin, former director of ReefQuest Centre for Shark Research in Vancouver, however, points out that it is not that simple. White sharks approach different prey in different ways. For example, how a white shark hunts and attacks a seal is radically different than how she approaches humans. "I spent five years in South Africa and observed over 1,000 predatory attacks on sea lions by great whites [and the] sharks would rocket to the surface and pulverize their prey with incredible force," Martin explains. In contrast, sharks typically approach swimmers and surfers with what Martin describes as "leisurely or undramatic behavior." This suggests that white sharks do not confuse people for seals.[25]

Indeed, sharks attack objects that bear no or little resemblance to seals. Collier describes one incident: "One morning off the coast of Oregon, on 24 August 1976, Mike Shook had been surfing about 45 minutes and was lying on his board attempting to catch a wave when he felt a slight bump to the rear of the board. This was quickly followed by a more forceful jolt, which caused the surfer to glance back. Much to his surprise, a large white shark had the rear of his surfboard in its mouth. The shark swam along the surface, pushing the board and rider five to ten meters before the end of the board broke off in the shark's mouth. The shark released the piece of board it had broken off, then submerged out of sight. Shook swam toward the jetty with the broken end of his board attached to his ankle leash."[26]

Collier is quick to underscore that white shark bites and attacks are not whimsical nor are they careless. Any encounter that ends in contact is risky and injuries, even seemingly minor ones, can be fatal. "When you really look at what a white is capable of doing given their musculature and strength, you see that the majority of bites in human attacks are very controlled. While this is hardly the experience of the person on the receiving end, the shark is actually giving the equivalent of a controlled nibble. He is curious and checking out something that is unfamiliar. They like to bump with their snout." Echoing Martin, he adds, "Again, in most cases, their bite on a surfer or swimmer is nothing compared with the determined attack when they strike a pinniped."[27] Attacks on swimmers and surfers usually end with a "bite and spit," where the shark chomps down but does not proceed to gnaw and eat. When the attack is fatal, it is usually due to exsanguination.

It has also been suggested that a bite is the shark's way of testing whether the prey is good enough, that is, has enough fat on his or her bones to warrant the effort and endangerment. Given that any encounter with prey can backfire and cause injury, a shark has to be sure that costs and risk are adequately compensated for by the fuel the prey can provide. The elevated temperature of the six-meter, two-ton shark body requires a constant source of energy-rich foods. Big prey is good, but fat prey is better. Generally blubbery seals, sea lions, and the occasional whale seem to fill the bill, while human ocean-goers, despite their readiness as easy prey, and the proximity of white sharks in need of fuel, do not. Humans' low fat-to-other tissue ratio may fail to satisfy the energetic needs of the shark in ways that a seal can.

Shark researcher Peter Klimley set out to test the fat criterion hypothesis. Using a rod and reel to troll variously pig, seal, and sheep carcasses outside California's Farallon Islands, he discovered that white sharks bit into all three species but consumed only the pig and seal. After the first chomp, sharks left the sheep alone. Klimley concluded that the first taste of sheep proved too fat-poor to be worthwhile.[28] There may be another reason that the sharks did not bother with the sheep carcass. In contrast to pumas who have mastered ways to efficiently skin sheep, white sharks may find dense lanolin wool and sea otter (*Enhydra lutris*) pelts too unsavory and too much trouble.

White shark dentition offers yet another intriguing clue. Shark teeth may serve not only as weapons, but also as sensors. A white shark has twenty-four to twenty-six teeth in the upper jaw and twenty-two to twenty-four teeth in the lower jaw. Behind this row of so-called functional teeth—those the shark is using at the moment—are six replacement rows, which give a white shark a total mouthful, on average, of three hundred and fifty teeth. Juvenile sharks replace their teeth every seven to ten days, whereas adults appear to replace their set every two weeks. When shark teeth fall out, those behind move up, escalator-style, to take their place. This guarantees that a shark is always equipped with sharp, ready-to-do-business teeth. The number of teeth dropped over a single lifetime adds up to thirty thousand or more, which, given that sharks have been swimming around the planet for hundreds of millions of years, may account for the great number of fossilized shark teeth that litter the strata.[29]

There is something more about white shark teeth. Their functional teeth can move forward and backward even while firmly embedded in their jaws. To shark biologist Rick Martin, this suggested an interesting hypothesis: perhaps this flexing ability communicates, via impulses to the shark brain, information about the prey at the end of a bite—in particular, whether or not the object is worthy of being a meal. Collier recalls a "3000 frames per second film clip showing a shark attack a carpeted target. It shows that it only takes the white shark 0.1–0.2 seconds between the time she grabs the target and when she releases it. In that infinitesimal time period, she has made the determination that she does not want the object. She bites then swims away. So, I suspect, given what Rick suggested, that more than fat content is involved in shark decision making."[30] (The footage also provides evidence that falsifies the image of sharks as mindless, instinct-driven machines. It takes astounding intelligence and aplomb to acquire data, process them, perform a cost-benefit analysis, and make a judicious decision, particularly when accomplished in less than 0.2 seconds.)

White shark diet varies with habitat, age, and individual preferences.[31] As a veteran of fifty-three years of studying and performing autopsies on sharks, Collier has substantive experience to back up his assertion that "if it were fat alone that triggered the feeding response then we would not see the kinds of species diversity in white shark stomachs such as bat rays (*Myliobatis californica*), Humboldt squid (*Dosidicus gigas*), and the hundreds of salmon-eye lenses that we found in the gut of a beached white shark on Ketchikan, Alaska. Further, I think the reason that so few humans are eaten, let alone attacked, is because we did not evolve as natural prey for white sharks. Our biochemistry, musculature, and so on, is just not part of white shark dietary evolution. Surfer report after surfer report describes how white sharks often just swim up, roll an eye at the person on the board, and then dip down and swim away."[32]

Basically, Collier continues, "attacks are based on two main factors: availability and capability. Availability means the kind of possible prey that comes along. Predators are opportunistic and if they come across prey that is 'good enough,' then they will go for it. But deciding whether to attack a given prey also depends on the individual's capability. Northern elephant seals (*Mirounga angustirrostris*) are wonderful white shark food, but not for a juvenile who is only five feet in length and who, in all probability, could not take on a 750-pound powerful prey without getting injured or killed but definitely could as a full-sized adult. That is why I say, animal diets don't shift, but *expand*."[33]

Elephant seals are formidable prey because they are fast, deep divers, masterful swimmers, and big, which together require a bit more strategizing than does going after the smaller harbor seal or sea lion. White sharks do use the tactic of attacking elephant seals from below, but after effecting a penetrating bite, they leave with the hope that the seal will bleed out and make feeding easy and safe. When the seal finally weakens or dies, the shark returns and carries his prize down lower in the water column, where he tears off and swallows hunks of flesh.[34] This method was confirmed in a 2015 study off the island of Guadalupe where researchers were testing the novel autonomous underwater vehicle (AUV). It was then that they made the first observation of subsurface white-shark predatory behavior. White sharks were "observed approaching, bumping and biting the AUV at depths ranging from 53 to 90m, thereby providing direct evidence of *C. carcharias* predatory behaviour at depth." In addition, the scientists saw the dead body of an adult, female Northern elephant seal floating on the sea surface. She had sustained four to five bites but "one of them was in the lower jaw and throat. This is a fatal bite because the shark's bite cut important arteries and veins from the seal. It is our belief that this brings about a quick death from rapid loss of blood."[35] Keep in mind, too, that the bite force of a white shark is regarded as "the highest known for any living species," calculated at 4,095 pounds (18,216 N).

Some may maintain that crocodiles hold this title, but even the most powerful, the saltwater crocodile (*Crocodylus porosus*), executes a bite force of "only" 1,302 pounds (5,792 N).[36]

So it seems that, like other carnivores, white sharks have evolved multiple sophisticated methods for tracking and procuring prey that draw on several detection modalities. This includes hunting at dawn and dusk when light penetration of water is at a maximum and what seems to be the use of scent to detect and direct their pursuit. In addition to excellent photopic vision (vision under well-lit conditions), sharks have lateral lines that allow them to zero in exactly to where prey can be found. Lateral lines are embedded in the face and sides of a fish. They are an elegant design, comprised of specialized epithelial cells called neuromasts that alert the shark to the slightest change in water pressure. The hair-like neuromasts, similar to those found in mammalian auditory systems, are covered with cupula, a gel-like substance that mechanically detects distortions in water pressure that are then translated via transduction into a neurochemical signal. These highly sensitive detection capacities give sharks an early warning advantage. Yet despite all these amazing evolutionary protective and predatory gadgets, white sharks are burdened with what seems to be a strange vulnerability—tonic immobility—that is, a shark can be induced into a reversible, coma-like state. At present, the mechanism that produces the stressful trance is unknown, but the phenomenon is obviously known to orcas.[37]

In 1997, on a crisp blue October morning in northern California, the whale-watching boat *Superfish* slid expectantly under the Golden Gate Bridge into the open waters of the Pacific. Crew and passengers were on their way to look for humpback whales that frequent these waters on their migration south. At about 10:00 a.m., a commercial fishing boat radioed the skipper, Mick Menigoz, to tell him that two orcas had been spotted off Southeast Farallon Island. The sighting was special. In the twenty-three years from 1973 to 1996, the big black and white whales had been spotted only occasionally. Captain Mick immediately kicked the whale-watching boat into gear and sped toward the spot where the orcas were last seen.

Superfish passengers were rewarded for this effort. The two female whales had been busy catching and eating a sea lion and passengers could see the arched forms of porpoising orcas just off the port side. While others were excitedly talking and watching the whales, fellow passenger Mary Jane Schramm, a naturalist with the Farallon National Marine Sanctuary, glanced down and spied a dark form about ten feet long swimming beside the boat. It was a white shark. In itself, the sighting was not unusual. Autumn is a prime time for seeing white sharks in the icy cold waters off San Francisco. What was unusual was what happened next.

Schramm watched as the shark took a forty-five-degree turn away from the boat. Then one of the orcas turned and made a beeline straight toward the shark. "There was a splash," recounted Captain Menigoz, "and then nothing." Suddenly, there was something. The towering black dorsal fin of an orca appeared and then, absent any thrashing or blood, the killer whale started "coming back toward the boat, carrying the shark in its mouth."[38] The smaller of the two orcas had the white shark by the neck and was towing the body upside down. The shark's mouth and ventral underbelly were exposed throughout this time. No blood or bite marks were apparent.

The orca continued to swim with the supine shark between her jaws. This went on for about fifteen minutes. Eventually a slick in the water appeared, indicating that the shark had been at least partially eaten. Western and California gulls, sooty shearwaters, and northern fulmars confirmed the kill, circling and swooping down to scavenge pieces of shark flesh. The orcas seemed interested only in eating the shark liver and left thereafter. Later, biologists identified the two orcas as CA2 and CA6, two females from an eccentric southern California pod that ranges from Monterey Bay to Baja.[39]

It is not unheard of for orcas to kill sharks, but it is rare. The Farallon incident was significant for this reason and for the way in which the orcas had subdued and killed the white shark. By inverting the shark on his back, the orcas had induced tonic immobility. According to Sam Gruber, a shark biologist credited with the discovery of the phenomenon, rotating a shark to a supine position causes a trancelike state. The fish's visual and gravity fields are reversed, which then causes a kind of sensory overload and triggers the brain to secrete serotonin, a neurotransmitter correlated with tonic immobility. The condition is not without costs because it exerts significant stress.[40]

Other species can experience thanatosis, including possums who "play dead," rabbits, various insects, and lizards. Because it can, as in the white shark's case, make an individual more vulnerable and because the state is physiologically stressful, scientists wonder how such a mechanism can be evolutionarily adaptive. There are three hypotheses for how and why tonic immobility is employed. A seemingly dead prey may deter predators who only capture live prey. Tonic immobility might buy time for escape if the predator likes to stash away his meal for later dining or if he prefers to carry the body away to eat somewhere else. (But biologists question how predators can be fooled because, as in the case of possums, the posture induced by tonic immobility does not resemble the dead version.) Some speculate that tonic immobility in sharks facilitates mating, though for most other species, nature has led to more co-participatory approaches for coupling. Tonic immobility remains one of the most "enigmatic areas of elasmobranch biology."[41]

It is fascinating to realize that orcas know enough about shark physiology to effect a successful attack on an extremely able opponent.[42] After the incident was analyzed and the orcas identified, there was speculation that the killer team were mother and daughter and that the mother was passing on to her protégée valuable cultural information: how to subdue a formidable adversary-cum-prey. The incident also underscored that although great white sharks may be the top of their game in the fish world, in the bigger pond of mammalian predators, orcas are number one.

Another new twist to white shark portraiture is that they are not the loners they are made out to be. Again, the veiled mystery of the ocean has misguided human perception. While the ocean may look fairly homogenous and unbroken to the onshore eye, it is as dynamic and varied as any terrestrial ecosystem. Thermal gradients, climatically shaped and driven currents, and prey patterns synergistically create a busy and complex world in which sharks live. Their social structures and processes mirror this heterogeneity.[43]

Shark sociality is a relatively new concept, but researchers are finding out that sharks, at least some of them, are much more civically minded than typically described. Researchers at an Ocean Science meeting in 2016 reported that, by attaching and monitoring acoustic tags on three hundred sand sharks (also called grey nurse sharks, or *Carcharias taurus*), they discovered that the sharks were quite social after all, enjoying numerous encounters in summer and perhaps other seasons.[44] Another team that is at the cutting edge of shark studies, led by Mauricio Hoyos Padilla, director of the marine research and conservation organization Pelagios Kakunjá, Mexico—with collaborators Fred Buyle, Yannis Papastomatiou, William Winram, and Lukas Muller—is finding similar social tendencies in white sharks.

White sharks routinely travel thousands of miles away from their well-stocked seal and sea lion food sources and gather with conspecifics in the middle of the Pacific Ocean. So far, these oceanic crossings do not seem to be prompted by any alternative or enticing food source. Unlike the seal-rookery-choked islands such as the Farallons and Guadalupe, where white sharks gather seasonally, these mid-Pacific "Shared Offshore Focal Areas" (SOFA; also referred to as Shared Offshore Foraging Areas) do not appear to offer any obvious buffets.[45] The mid-sea space between Hawaii and Guadalupe, Mexico, is instead regarded as somewhat bereft of the luscious foodstuffs that sustain white sharks. Smaller prey may be there to tide the sharks over, but it remains a puzzle as to why an eating machine who requires so much energy input would intentionally leave to go where sustenance seems poor.

Females swim around within a 250-kilometer radius of the SOFA, while males tend to focus on a core region. Sojourns can take up to six months. In the summer, males sometimes return to their island food magnets, leaving females behind who hang around for another year. Female and male white sharks aggregate, but without any obvious signs of physical interaction. This suggests that the space functions like a kind of underwater *place de rendezvous,* where a shark can taste a few dainty morsels, have a briny drink, and meet up with new and old friends. Scientists have dubbed it the White Shark Café.

Aerial surveys and photo identification reveal an entire repertoire of shark social behaviors—purposively swimming near each other, side-by-side swimming, and even engaging in "splash fights."[46] But the significance of the communiqués remains obscure. Unsurprisingly, captive studies and network analyses demonstrate that sharks show social preferences—sharks care for each other. This provides an important reminder, namely that "species often considered solitary might in fact integrate some aspect of social interaction into their behavioural repertoire."[47] Sociality can take many expressions and patterns.

Using novel methods of tagging, as well as transceiver devices to triangulate locations, Hoyos Padilla and company monitor sharks from August to December to identify predation tactics while white sharks are close to seal colonies in Guadalupe. Then, in January, they move on to White Shark Café. This spatiotemporally explicit approach allows for analyses of sharks' social networks that may provide insights into the area's popularity and function and should form a solid basis for building an understanding of shark social psychology.

Already, researchers are discovering all sorts of fascinating tidbits. For example, one white shark in the Mexican study, Deep Blue, is over twenty feet long and quite the matriarch. Her age is estimated to be over fifty years. At the time of her tagging off Guadalupe Island in June 2015, she was heavy with shark young. Hoyos Padilla, who has studied individual and groups of white sharks in detail, attests that videos of Deep Blue in close approach show an incredibly tender, soft expression and gentleness, something that Charles Darwin would be likely to claim as further evidence of his own studies showing that all humans and other animals exhibit emotion with remarkably similar expressions.[48] Naysayers and adherents to the shark-as-unfeeling-machine paradigm may shake their heads in disbelief. But Hoyos Padilla's description, rendered precise and insightful by scientific rigor, evokes an image so powerful that it breaks through the conventional monster shark image. In its place, what emerges is the glow of motherhood beaming from the face of a cold-blooded fish—a true lesson of "one can't tell a book by its cover."[49] Furthermore, Hoyos Padilla's observations confirm what neuroscientists have predicted: vertebrates, even sharks who seem so very different in so many ways from us humans, have the same evolutionarily crafted internal neural circuitry, which expresses itself in subtle shades of a wide range of emotional and behavioral responses.

Site fidelity is both a strength and a vulnerability—a strength because it obviously plays a key role in shark society, and a vulnerability because it makes it easy for trophy hunters and fishermen to find, catch, and kill the sharks during what is probably a very susceptible period. One thing is clear: white sharks are not "lone rangers" passing through the deep blue anonymously. Their predictable migration paths demonstrate that they are intimately connected in ways and at spatial and temporal scales that researchers are only just beginning to understand.[50] White sharks off Australia exhibit a "high degree of residency to specific aggregation sites, to which they return regularly over multiple years."[51] It also seems that white sharks may collect and function as "clans" by following a fairly strict timetable, coming together in a kind of shark "time share."

Ralph Collier describes how, starting in 2008, white shark attacks off Surf Beach, near Santa Barbara, California, have occurred in regular two-year cycles. "In 2008, a surfer was attacked at Surf Beach, but was uninjured. In the last second, he lifted his legs and the shark missed and bit the back of his surfboard. Two years later, in 2010, surfer Lucas Ransom was attacked by a white shark: his leg was torn off and he died. Then again, in 2012, thirty-nine-year-old surfer Francisco Javier Solorio Jr. was killed at the same location, and again, in 2014, two kayaks in the same areas were attacked in succession by a shark more than nineteen-and-a-half feet in length. So it may be that there is a clan with a two-year migratory cycle."[52] Another attack occurred, two years later again, in October 2016. Marine scientist Marc Aquino Baleytó, who works with Hoyos Padilla, adds that the attacks may "correspond to the two-year cycle of pregnancy of the females (12–18 months) . . . [and] it may be that these attacks were made by females who come close to the shore to give birth.[53]

This expanded understanding of white sharks certainly shifts the commonly held perspective of who these wonders really are. This is no surprise for Fred Buyle, who has known white sharks within the intimate space of only a meter or two. Any collective myth was, in his mind, debunked in the first moments of an initial meeting with white sharks off South Africa:

> Back in 2007, I met up with a friend and colleague in South Africa with the specific purpose of meeting a white shark. It first required that we take a boat and travel two hours to a remote spot off shore. I trusted my friend completely. He was very canny

and experienced so I felt relaxed and happily excited at the prospect of finally meeting the infamous white shark.

Even though the area was supposed to be "shark infested" and we had put out bait, it took nearly four hours before the first shark showed up. This was totally a surprise to me because everyone said that given the first drop of blood, sharks immediately rush in and begin to circle. Over the years, I have come to know this is false. A lot of times, you can spend days waiting and looking for sharks and not one shows up.

Suddenly, a white materialized not more than fifteen meters away. He wasn't that big, only three meters in length, but even if whites are on the short side, their girth makes up for it. No matter what size, whites look enormous. They look and are twice as big as a tiger shark of the same length. But even though I felt relaxed, for a split second the music from *Jaws* started up in my head. The shark turned towards us and aimed straight at me and I thought, "I must be out of my mind—what the hell am I doing?!!" But then, as he got close to me, I saw that he was quite curious. He passed about three meters away and I saw in his eyes that he was quite shy. Wow, that was a revelation. Here he was, a top predator of the ocean and he was as uncertain and nervous as me! He spent almost thirty minutes swimming around us before he decided to move on. I was floored that he just stuck around for so long. He may have been thinking, "Hm-m, maybe an easy meal," but none of his body behavior showed that, and as an apex predator he knows the exact angle to make the perfect strike. In comparison to a seal or a shark, we humans are very clumsy and so if he had decided to attack, there would have been no way we could have escaped. With a blink of an eye, he could have easily cut us in half. But he did not take advantage of us in the water.

It seems like a paradox because we assume that white sharks are rulers of the sea and that they have only one thing in mind: to track and eat you. But that day, I learned something different and it was a big kick in the ass lesson. I realized that despite their

> incredible strength and intelligence, sharks, unlike most humans, are very restrained. Sharks don't attack or kill on a whim. They don't abuse their power.[54]

Since that fateful day, Buyle has spent hours free-diving around and with white sharks. In retrospect, when thinking back on his encounter with the young male and the photos he took, Fred is struck by what he has learned over subsequent observations with many other white sharks. In contrast to most young sharks, this one lacked any scars. Male sharks grow up in the rough-and-tumble, risk-and-dare style of restless youngsters. When white sharks gather, the largest, usually older females, swim at the top with subordinates ranked down the water column. When this social order is challenged or violated, disapproval is communicated through sharp nips and cuts. An attack is usually preceded by an exhibition of impressive body language including depression of the pectoral fins, parallel swimming, head shaking, and gaping.[55] By the time they are adults, shark bodies are laced with marks and slices from bigger, more prestigious elders.

That Buyle's shark had retained his silvery smoothness suggests that he deviated from the usual wild-bunch profile. Somehow, whether by avoiding shark rumbles or by deft escapes, he had not fought with conspecifics. He showed a different personality than what has been typically observed in other white sharks. The idea of a fish or other wild animal personality is not far-fetched. Neuroscientists have shown that sharks' brain structures and processes are common to mammalian vertebrates including orcas, grizzlies, and humans, so it is logical and scientifically consistent that our finned friends will exhibit a diversity of temperaments. Details of their interior landscapes may differ from our own, but we share a common psychological geography.

Fish do have a very different rearing environment than mammals and other altricial species (those whose infants require feeding and care by parents or relatives before they are able to stand on their own four feet or two wings). This means that the influences shaping shark minds are also different. Like birds, sharks experience a variation on a vertebrate developmental theme, but most of the wheres and hows of shark reproductive life remain obscure. What information has been gathered comes from serendipitous discoveries

The enigmatic Hammerhead Shark. Photo credit: Fred Buyle.

from bycatch or other fatal events that reveal in utero embryos and maternal body conditions and chemistry. Based on these piecemeal data, here is what most researchers agree on with respect to white shark development.

Gestation is estimated to take somewhere between eleven to eighteen months, with birthing occurring in spring or summer. White sharks are aplacental viviparous, which means that their young depend on an egg yolk while occupying their mother's womb. Other shark species are either oviparous, laying eggs that are delightfully called "mermaids' purses," or, like bull and hammerhead sharks, viviparous—a more mammal-like way of development that involves a placenta and umbilical cord for nourishment of the embryo.

Budding white sharks get their sustenance during gestation by feeding off unfertilized or unhatched eggs, a cannibalistic oophagy that some interpret as foreshadowing white sharks' mythic adulthood ferociousness. When baby sharks reach about four to five feet in length, they are ejected face-first into the blue where they must make it on their own. (By contrast, many species exit their mothers tail-first to avoid drowning.) Others believe that newborn shark length in white sharks is, similar to other species, including thresher sharks (*Alopias vulpinus*), not fixed but dependent on the size of the mother. The bigger the mother, the bigger her young. In the 1970s, a commercial fisherman caught a young white shark in his gill nets used to catch California halibut (*Paralichthys californicus*). The baby shark was only thirty-two-and-a-half inches and still showed an umbilicus scar between his pectoral fins.[56]

In principle, then, white sharks' brain-mind development follows that of other vertebrates, at least to some extent. Psychological tuning in mammals takes place within close relationships from the get-go. For example, infant elephants drop from their mother's womb into a constellation of aunts, siblings, and cousins. This microcosm of pachyderm society is the lifetime residence for females, whereas young males will leave the natal developmental context to embrace a second comprised of elder bull mentors. These young depend on their parent and others to provide food and protection for a period of time after birth.

But life in an egg and immediate separation from the mother shark seem radically different. Baby sharks do not have to rely on someone else to provide food and protection. Precocial young—those who pop out of an egg ready to fend on their own—seem to have very little mammalian-like tutelage once they leave their egg, or, in the case of sharks, the sheltered interior of their mother's belly. But closer examination reveals that egg-laying birds and fish are still able to experience relationally based development.

Biologists used to draw a hard and fast line between precocial birds, such as chickens, and altricial birds, such as parrots. But "behavioural development in all parts of the altricial-precocial spectrum is more flexible than originally thought."[57] Avian developmental pathways are often a mix of altricial and precocial traits. Furthermore, while white sharks may not feed their young, epigenetic studies suggest that not everything an egg-born infant knows is passively inherited. With shell or without, an embryo in utero is influenced by his or her environment, internally within the shell or mother, and externally as the parent tends the egg(s) and then the infant animal.

Studies on chickens show that epigenetic modifications occur from the start. The mother's "environmental signature" is passed to her chick, what biologists refer to as "F1," through egg content, but also to the successive generation (F2), "since the developing offspring bears the primordial germ cells (PGC) that later differentiate into gamete precursor cells and finally lead to the individuals of the F2 generation."[58] Neuroendocrinal research over the past decade has shown just how important these exchanges are.

In 2001, Michael Meaney, a professor of biological psychiatry at McGill University, published a ground-breaking paper showing the intimate relationships between inside and outside, inheritance and experience.[59] In a series of gene-environment interaction (G × E) experiments, Meaney demonstrated that an infant's early experiences significantly affect neural development and function via modifications to the genome that do not change nucleotide sequencing. This new concept (and field) was called epigenetics (or neuroepigenetics), and it provides insights into how environmental differences interact with genes to produce individual phenotypic differences in appearance, thoughts, emotions, and personality. Or as Meaney summarizes: "Genomic variation at the level of nucleotide sequence is associated with individual differences in personality and predicts vulnerability and resistance to a wide range of chronic illness."[60] And it all happens in the tender context of maternal care.

With the discovery of epigenetic effects, the age-old nature versus nurture debate was dispelled by establishing a causal relationship between how infants are reared and how their neuroendocrine stress response is programmed. Essentially, the parent or other caregiver functions as a socio-ecological broker for the baby. The neurobiological and psychological state of a mother and her interactions with her young modify the gene expression responsible for regulating behavioral and neuroendocrinal responses to stress. Caring, appropriate parenting and good health lead to progeny who can handle the stress of a dynamic ecosystem. Poor-quality care and poor health tend to diminish the ability to respond well to environmental ups and downs. It is the combination of early experiences in the womb and the broader environment that appears to set the path along which a child, cub, fry, or chick journeys and who they turn out to be.[61]

Before recent explorations into the neurobiological mechanisms that account for the wide differences in how people act and think, scholars and pundits of the past were coming up with their own hypotheses. Personality theory is an old psychological standby that dates back in the West to the Greeks, whose culture flourished starting around the fifth century B.C.E. Hippocrates, who is credited with being the progenitor of modern clinical medicine, proposed that personalities could be classified using two orthogonal axes—ranging from dry to moist on one hand and hot to cold on the other. The four quadrants that result correspond to the four humors—air, water, fire, and earth. These were used to explain variations in personality and moods of individuals. A similar natural-elements approach is found in many other cultures, including India's Ayurvedic tradition. But it was not until Franz Gall, a German neuroanatomist who lived from 1758 to 1828, that personality types were formally associated with specific locations and functions in the brain. The field he founded, phrenology, used measurements and brain geography to identify and diagnose psychopathology and criminal behavior. Because of its perceived dangerous political use and radical departure from empirical foundations, the field has been discredited. But Gall was basically on the right track when he related brain function to the structures of the brain.

Since both Hippocrates and Gall, there has been a virtual burgeoning of methods and tests to describe personality types. In its most general definition, personality is defined as the behavioral repeatability of individual traits over time and in various contexts. One current standard is known as the Big Five personality test, also referred to as the Five Factor Model (FFM). A set of questions forms the basis of a personality profile that assesses five categories (or factors): openness to experience, conscientiousness, extraversion, agreeableness, and neuroticism. Each factor ranges across a continuum.[62]

Although the Big Five was designed for humans, it has been deployed for a diversity of species. The increasingly overlapping interests and efforts of those working toward animal rights, which tends to focus on individual wellbeing, and those working on conservation, which seeks to preserve population and species integrity, have turned attention to understanding wildlife from the inside out. It is estimated that up to 30 percent of a population's behavioral variability can be attributed to individual personality differences. Furthermore, multiple phenotypes, including those of behavior, may arise from a single genotype, and different genetic backgrounds have different propensities for phenotypic variability.[63] How and who we and our animal kin become depend on a complex of interactions.

University of Texas psychology professor Sam Gosling is a leading researcher in the systematic investigation and formalization of animal personality research. Personality traits have been identified in no fewer than sixty-four different species ranging from octopi to mice, chimpanzees, and goldfish.[64] Fish biologist David Jacoby from Australia has extended the work, which has focused mainly on mammals and birds, to probe nuanced differences in various fish, including sharks.[65] Critics have argued that these data are suspect because of observer variability and anthropomorphic projection and that sociocultural views can bias interpretation. In addition, the presence and variability of personalities do not seem to fit comfortably in conventional evolutionary schema.[66] But evidence supports the validity of a comparative approach to studying animal personalities.

Levels of agreement among observers are strong and predictions are consistent.[67] When individuals independently evaluate the personality of a dog, chimpanzee, or a member of another species, they agree within statistical bounds. Not only are expressions shared among species as Darwin pointed out, affective neuroscience research demonstrates that the internal structure of the FFM's psychometric measures is grounded in primal subcortical emotional systems that are common to humans and other animals.[68] This relationship accounts for the success in empirical studies where human observers are able to successfully predict "real world outcomes," such as breeding patterns and immunological resilience, using subjective personality assessments.

Fred Buyle was asked to answer the questions in the Big Five personality test for several shark species to see how his perceptions relate to a more formalized assessment and in so doing, to obtain a glimpse into white, tiger, and other shark psychologies. After Buyle entered his responses, the automated test independently calculated scores for each of the five categories. In addition, they help us as bystanders see into the mind of a human observer who must be a perceptive psychologist himself if he is to avoid becoming prey to his subject.

White sharks scored low on openness to experience, agreeableness, and neuroticism. These are individuals who are conventional with narrow interests and often exhibit a harsh and critical nature, yet go about life in a calm and secure manner. They score high on conscientiousness, which is characteristic of careful, well-organized, and reliable individuals. This coincides with Fred's personal experience: "White sharks are kind of fragile. As apex hunters with a big energy budget to maintain, they have to be focused, with a one-track mind and the ability to perform efficiently and fast, consistently. Even if you're kingpin, hunting is dangerous so you can't afford to mess up. You have to plan well and keep your act together. A white shark is always on point. Look at the ways they hunt and feed—fast, direct, bite and swallow. I wouldn't say that white sharks are grumpy—even though they don't look very happy—it's just that their job forces them to be self-centered and in control. I think that's why they don't seem very friendly. Curious and calm, but pretty single-minded, which is why I make it a rule to never stay in the water with a white shark longer than forty-five to sixty minutes. More than that, you tend to get tired and stop paying attention. That is when accidents can happen. So you owe it to the shark to be present."[69]

In contrast, while also scoring low on openness to experience, more moderately on agreeableness, and very low in neuroticism, tiger sharks appear more relaxed and possess a more secure sense of self. Tiger sharks also scored low on conscientiousness, which typifies somewhat disorganized people. Again, personality test results coincide with Fred's description of what he has experienced: "Tiger sharks are very confident. They are strong and built like a tank. But, unlike whites, tigers can afford to be a little more at ease. They can get their food by scavenging off reefs and rocks. There is usually some piece of fish or other food around which does not require the intense hunting strategy of whites. So in some ways their job requires less risk and less focus. I think that this is why tigers are easier to be with in the water. You always, and I mean *always,* have to keep your eye on sharks no matter what the species. You have to pay attention. But in my experience, tigers like to hang with humans more than whites, for example, and even allow divers to gently push them away if they get too close. Sometimes, you will have seven or eight tigers around you and there is no tension, no problem, nothing bad happens. I would say that tigers look and act pretty happy most of the time."[70]

But personality theory and epigenetics caution that not all people or sharks are the same, as two incidents illustrate. In one instance, after her tiger-shark-feeding bucket-list vacation in Mexico, a French woman wrote a post-card to her beau saying, "I expected terror, I expected vicious teeth, and frenzied biting. Instead what I saw was *la tendresse,* tenderness, and eyes of the innocent staring at me." Elsewhere, thirteen-year-old surfer Bethany Meilani Hamilton was sitting on her board waiting for the right Kauai wave when a tiger shark, seemingly unprovoked, acted very differently than his or her counterpart in Mexican waters. Instead of "*la tendresse,*" the tiger shark, whether out of curiosity, hunger, or hostility, bit off the arm of the young American surfer. (This tragic event, however, has not deterred her and at twenty-five she is still involved in competitive surfing.[71])

While there are traits shared within a culture or species, just as in the case of humans, behind every pair of eyes lies a unique history and a unique set of circumstances that inform the moment of contact. Consequently, not all tiger sharks may have identical levels of a "secure sense of self." Although this quality is obviously necessary for survival—self-doubt in a hunter can lead to fatal distractions—just as there are differences in how individual humans view themselves and their abilities, so too with other animals. Ralph Collier also emphasizes the importance of recognizing shark individuality. "There are so many unanswered, important questions about sharks and their modes of living. For instance, last year, off Guadalupe, we saw a group of white sharks and there was one female who had copepods all over—between her pectoral fins, some around the pelvic areas, dorsal—but she was the only one who had them. Why is that?"[72]

Bull shark personalities revealed quite a different story. They are not your exemplar hail-fellow-well-met personality. Bull sharks' Big Five scores suggest they are neither agreeable nor open. Like the tiger, bull sharks score high in neuroticism. They also score high on conscientiousness. It is this combination that puts divers on their finned toes. Buyle comments, "I don't like to say anything negative about sharks because basically they are all pretty good guys. They never seem to do anything for no good reason. But I will say that bulls have a kind of schizophrenic bipolar personality. Sometimes they are quiet, kind of sneaky, then other times completely aggressive and crazy."[73]

In this context, the personality of the young male white shark whom Buyle had met in 2007 stands out. He showed a sensitive curiosity and positive attitude to the human with whom he came face to face. His score for openness and agreeableness was higher than what might be considered average for a white shark. Sitting back with a fond smile and sigh, Buyle now considers that the young shark was "very white shark like, but different," "a cool guy—open, not aggressive, not pushy, just someone a little shy with a curious, and I might even say, thoughtful, personality."[74]

All of this may sound fanciful, but it is imperative to be able to read the personality of a shark or any other individual who can make a life or death choice for you. Charlie Russell agrees that the most important thing to keep in mind when around a grizzly is that individual's personality: "Over the half century of living side by side with grizzlies, I have learned a lot of things, but one in particular stands out—differences between individual bears are much greater that those between species. A lot of the trouble people get into with bears starts when humans don't pay attention to whom it is exactly that they are meeting up with. If you see a grizzly just as any old grizzly, then you are setting yourself up for potential problems. Bears play by the rules, but like everyone else, each one has a particular personality and history and whatever is in that history will influence what a bear thinks and does—and it is up to humans to appreciate this. Every time you meet a bear, it's different, it's a unique relationship between the two of you."[75]

Marie Levine concurs: there are individual personalities within cultural and species personalities. Subsequently, while certain personality patterns characterize a given species, it is individual differences that are most important. "Every shark is an individual. Like dogs, cats and people, each has his or her own unique personality, and moods and emotions may differ from day to day. Some white sharks I've worked with tend to be very mellow, others are skittish and easily frightened, while others have been intensely curious. It would be misleading to ascribe a particular behavior to an entire species."[76]

Fred Buyle joins in with a similar opinion and adds an interesting view on how he interacts with and approaches sharks: "It is not just me and just him, it is us together at that intersection between the shark's psychology and my own. When I am in the water, swimming with sharks or whales, I don't feel it is my brain where all the processing goes on. It's that zone between me and the shark, which is the reality. When I am in that zone, I am present, enjoying the moment, and reading and connecting with that individual. In a lot of ways, I forget he or she is a shark. The only person who counts is that being in front or around me. When I am not in that zone, it is a completely different experience. It is as if I am not there and it is then that I can be stupid because I am not really listening in to who I am with. I think the shark senses that too. Definitely, I have seen it with the white sharks."[77]

The new convergence of epigenetic research and natural psychobiological history has huge implications for understanding and interpreting wildlife behavior, in particular, predicting trends in species' personalities. Personalities are phenotypic expressions of epigenetic variation and the vast majority of the Earth's "epi" has radically changed. While the Earth's oceans have been in dynamic flux, life for marine species is nothing like it has been in past centuries because of the arsenal of human-caused stressors, which include not only increases in water temperature and acidification, but also pollution, overfishing, mass killing, and dramatic imbalances in biodiversity. For sharks who have lived on the planet in one form or another for 400 million years and managed to coevolve successfully with their environment throughout this time, present conditions represent a significant departure from established, albeit dynamic, patterns. Modern human ecological disturbances are all press and no pulse. The extreme pressures exerted on the planet's ecosystems leave little wiggle room for adaptation. And when a shark mother experiences trauma, poor nutrition, pollution, or other negative stressors, the effects are neurobiologically transmitted to both placental and aplacental embryos, just as those effects would be transmitted to a human embryo and fetus. Everyone paying attention has seen the change.

Former businessman and avid ocean fisherman Chris Fischer now has taken up white sharks. He has noticed the change in shark behavior over time. It has been more than forty years since the film and book *Jaws,* and "every shark since has been shot, hooked, or whacked, so now what we have is a very cautious fish. Why wouldn't they be afraid of boats? That's my theory. I can just throw that out. I'm not a brainiac. I'm a fisherman."[78] He maintains that white sharks are circumspect, they "grip the bait, and they swim away with it" and, if "they feel any tension—*thwoo!*—they spit it out. They don't have any tension like that in the wild." Everyone assumes that these huge apex predators are "going to come in and hit the bait," but instead "they swim up and extend their teeth and delicately touch it. We're dealing with the most nervous of the most nervous of the species that survived two hundred years of whaling and long-lining, and now we have their offspring."[79] Fischer's statement is hauntingly mirrored in tribal human cultures such as the Inuit, whose "elders in the community were brought up in circumstances profoundly different than those of their children and grandchildren."[80]

Change has been observed elsewhere in the oceanic trophic web. Starting in the mid-1990s, Cornish marine researchers began to see battered, dead bodies of harbor porpoises. At first, it was thought that they had been killed by shock waves from U.S. Navy exercises or by air guns used by Scottish oil-rig technicians to detect undersea caverns, but postmortems showed that the injuries were quite specific and targeted. The porpoises had "broken ribs, imploded lungs, damaged livers and massive internal bleeding" symptomatic of "prolonged, focused attacks."[81] Then the cause of death was discovered: bottlenose dolphins (*Tursiops truncatus*) were observed hunting down and killing porpoises. Even more disconcerting was the realization that the dolphins had employed their sophisticated sonar system to target vulnerable parts of porpoise bodies to better their aim when ramming. Porpoise killings were in no way accidental but, from a prosecuting attorney's perspective, first-degree murder.

The suddenness and seeming break with normative behavior stunned researchers. A number of explanations have been proposed. One study seemed to indicate food competition and territoriality. But while the diet of the two species overlaps, they occupy different food niches. Porpoises like to eat shoaling fish and squid in cold waters. Dolphins do too, but are a bit more catholic in their dietary tastes and areas in which they sup. In other instances of "porpicide," there did not appear to be any shortage of prey for either species. Another hypothesis suggests that since porpoises are about the same size as dolphin calves, the cross-species killings might be related to dolphin infanticide. Five out of eight dolphin calves found stranded had similar injuries. Dolphins had been sighted "playing" with dead infants much the same way they had with the porpoises they had pursued.

In 2007, on the other side of the world in the Pacific Ocean, a "gang" of young male dolphins chased at high speed, rammed, then drowned lone porpoises.[82] The onset of the attacks was precipitous and dramatic. In six years, from 2005 to 2011, forty-four harbor porpoises along the California coast died from attacks by young male dolphins. The coincidence of attacks with the height of breeding season may indicate that high levels of testosterone had aggravated male actions.[83]

But what cannot be ruled out are environmental stress and traumatic social disruption. Although the link may not be explicit, there is a strong correlation between killings and the epidemic appearance of epidermal disease, lesions, and discolorations among bottlenose dolphins. Common bottlenose dolphins live throughout the planet's oceans and are subjected to a broad spectrum of assaults that include chemical pollution, mass killing in culls, diminished numbers of prey from overfishing, habitat degradation, direct and indirect disruption from boat traffic, commercial dolphin watching and interaction tourism, as well as noise and marine construction.[84] All impose chronic stress and trauma on the species and are therefore candidates for effecting epigenetic pressure. This makes it difficult to determine whether certain behaviors are normative for the species or are the result of an epigenetic adaptation.

Researchers regard dolphin aggressive patterns as aberrant, but their intensity and widespread nature suggest that they may be a new norm. Certainly this is the case for young African elephant bulls, and now females, whose deviance from the pacific elephant norm is now the signal, not the noise. The parallels between homicidal dolphins and elephants are startling. Symptoms of post-traumatic stress disorder (PTSD) caused by widespread human violence is now epidemic in South Africa, where rhino-killing elephants were first observed. Not long ago, elephant bulls assaulted and killed over a hundred endangered white and black rhinoceroses. Their mental and emotional states at the time of the killing were also correlated with sexual aggression. In all cases, psychological disorders of intra- and interspecific killing, infant neglect, and infanticide trace to a series of developmental disruptions. Although their "societies" may not be as structured as eusocial species are, sharks may be changing too. Shark brains, fish brains, dolphin brains, elephant brains, and human brains run on a common neurobiological engine that is responsible for regulating stress and is vulnerable to environmental pressures.

Subsequently, while sharks may be formidable hunters, they are unable to counter the persistent onslaught by humans. There are varying estimates of white shark population sizes, but all agree that white sharks, and sharks on the whole, are under severe pressure from legal and illegal fishing.[85] There has been a global "boom" in commercial fisheries' shark catches, as they seek to obtain, in addition to shark flesh, liver oil, and cartilage, shark fins that are used in shark-fin soup and for other products. Finning, which entails catching sharks, pulling them on board, slicing off their fins, then tossing them back into the water, is one of the most serious threats. The practice has not only had a huge negative impact on diverse shark species; it also causes great suffering for the sharks who, after being thrown back into the ocean finless, either suffocate or are eaten by predators.[86]

It is estimated that approximately 100 million sharks—or depending on whom you ask, between 63 million and 273 million—are killed each year in commercial fisheries.[87] Some species have declined by as much as 99 percent and the International Union for Conservation of Nature (IUCN) has announced that a quarter of shark and ray species are threatened with extinction. Nonetheless, protection for sharks remains minimal. As of 2015, only a few shark species, including the scalloped hammerhead (*Sphyrna lewini*), largetooth sawfish (*Pristis pristis*), and some other sawfish species are listed under the Endangered Species Act.[88]

Fred Buyle has also noted a shift: "You know how many white sharks are left? A handful worldwide relative to what there used to be and so many are killed every year. White sharks take up to fifteen years to mature and have babies. When you hear that and that every other shark, ray, and whale is having the same problem, it makes me think that we might be the last generation to be able to see and live with these amazing beings. It is terrible to think what we have done to these wonderful, ancient cultures. I am not worried about the planet. The Earth has made it through other extinctions. But I am worried about us, our species, for what we have done and what will happen to us once we are all alone on a planet that doesn't like us anymore."[89]

2

Grizzly Bears: How Brains and Minds Develop

She makes a decisive clawed grab and dives into the icy water, sending her right arm arcing around the stunned body. He struggles to regain consciousness, trying to reorient, but just as he draws a breath, a second assault renders him senseless. The polar bear pulls the beluga out of the water and drags the

huge body onto the floe. With a precision bite, she kills the whale and begins to tear into the blue-white flesh.

Polar bears (*Ursus maritimus*) are usually considered the largest species in the bear world. When these arctic gods hunt and kill a seal or beluga twice their length and double in weight, there is little doubt that they reign supreme in their icy kingdom.[1] But a grizzly bear's strength is about on par with that of his polar counterpart. The Kodiak version, *Ursus arctos middendorffi,* who lives on islands off Alaska, is massive by any calculation. Grizzlies are as mighty as their snow-colored cousins—they just wear their strength and size differently.

The grizzlies' reputation for fierceness is legend. Even the distant cool of science cannot remain unmoved. "The grizzly bear inspires fear, awe, and respect in humans to a degree unmatched by any other mammal in North America . . . Contributing to the aura of the grizzly bear is a mixture of myth and reality, ferocity, unpredictable disposition, large size, strength, huge canines, long claws, keen sense, swiftness, and playfulness."[2]

If their mythic stories are whispered around campfires more frequently than those of polar bears, it is simply because encounters between grizzlies (*Ursus arctos ssp.*) and humans have been much more plentiful. Although today grizzlies range into the frigid climes of Russia and Canada, historically they also roamed much farther south, into Texas and Mexico, and as far east as the Great Lakes. Think of it—grizzlies ambling along golden cliffs overlooking the glittering Pacific. But the fact that grizzlies favored this territory has led to unfortunate consequences. Grizzly habitat is coveted human habitat—rich fertile valleys for agriculture and livestock, streams stocked with salmon, and hillslopes bejeweled with berries and mantled by timber. The Euro-American imagination was topped only by the colonists' greed.

Unlike American Indians, settlers were unwilling to share the land with bears. The unquestioned slaughter of bears was rationalized by the claim that bears are dangerous to humans and to their pocketbooks. Although the potential for calf predation was minimal, bears were killed routinely. The consequences of this mass killing were grim. In a blink, within a hundred years,

(Overleaf) Fisherman, Brown bear, Kamchatka. Photo credit: Charlie Russell.

by the 1930s, over 98 percent of grizzly homelands in the lower forty-eight states had been appropriated, and the great bear had been beaten back to tiny pockets of refugia. Today, grizzly populations are a mere fraction of their nineteenth-century numbers. Only an estimated 800 to 1,020 survive south of Canada.[3]

The predominant cause of grizzly deaths remains humans. While statistics vary by region, up to 90 percent of the mortalities result from hunting or from citizen kills related to poaching; from misidentification for the grizzly's cousin, the black bear (*Ursus americanus*); from claimed self-defense; and from "management" by government personnel.[4] The reasons for human hostility and intolerance are based on an image of grizzlies as aggressive, out-of-control monsters and a belief that the only way to deal with a bear is to fight fire with fire, or better yet, terror with terror.

Seasoned woodsmen, however, do not share this outlook on grizzly bears. Consider Enos Mills, who lived nearly forty years in grizzly country. The guiding force behind the establishment of Rocky Mountain National Park, he spent untold hours watching and collecting details about grizzlies. According to Mills, the inspiration for the subspecies' scientific moniker *horribilis* relates to an unfamiliarity with bears and with nature overall: "Is the grizzly bear ferocious? . . . Speaking from years of experience with him my answer is emphatically, 'No!' The majority of people who say he is not ferocious are those who have studied him without attempting to kill him; while the majority who say that he is ferocious are those who have killed or attempted to kill him."[5]

America's icon of natural history, John Muir, concurs. For nearly half a century, Muir trekked through a grizzly-laden wilderness—now barren of the species—and never carried a firearm. He chided city folk and journalists for their inaccuracies: "There are bears in the woods, but not in such numbers nor of such unspeakable ferocity as town-dwellers imagine, nor do bears spend their lives in going about the country like the devil, seeking whom they may devour."[6] Muir recalls an encounter with a bear that showed him the error of this collective myth. One day, out walking in California's Sierra Nevada Mountains, Muir came across a bear. He quickly hid behind a tree and after "studying his appearance as he stood at rest, I rushed toward him to frighten him, that I might study his gait in running. But, contrary to all I had heard about the shyness of bears, he did not run at all; and when I stopped short within a few steps of him, . . . he held his ground in a fighting attitude." Embarrassed in sight of the offended bear who stood in front of him, Muir saw that his mistake "was monstrously plain." "I was then put on my good behavior, and never afterward forgot the right manners of the wilderness."[7]

Of lesser standing among bear experts, the flamboyant frontiersman and poet Joaquin Miller describes his observations of the now-extinct California grizzly (*Ursus arctos californicus*) in the nineteenth century: "In the early 50's, I, myself saw the grizzlies feeding together in numbers under the trees, far up the Sacramento Valley, as tranquilly as a flock of sheep. A serene, dignified and very decent old beast was the full-grown grizzly as Fremont and others found him here at home. This king of the continent, who is quietly abdicating his throne, has never been understood. The grizzly was not only every inch a king, but he had, in his undisputed dominion, a pretty fair sense of justice. He was never a roaring lion. He was never a man-eater. He is indebted for his character for ferocity almost entirely to tradition, but, in some degree, to the female bear when seeking to protect her young." Miller witnessed firsthand how the great grizzly, who had lived so long and well with indigenous tribes, "went out as the American rifle came in."[8]

There are others who came to agree that bear reputations are built more on myth than fact.[9] In *A Sand County Almanac,* Aldo Leopold writes with wistful bitterness about the killing of "Old Bigfoot," the last of the big grizzlies of Escudilla, Arizona: "The bureau chief who sent the trapper was a biologist versed in the architecture of evolution, but he did not know that spires might be as important as cows. He did not foresee that within two decades the cow country would become tourist country, and as such have greater need of bears than of beefsteaks."[10] Leopold does not hold himself blameless. He understood the blindness of unwavering rectitude: "We forest officers, who acquiesced in the extinguishment of the bear, knew a local rancher who had plowed up a dagger engraved with the name of one of Coronado's captains. We spoke harshly of the Spaniards, who, in their zeal for gold and converts, had needlessly extinguished the native Indians. It did not occur to us that we, too, were the captains of an invasion too sure of its righteousness."[11]

Since 1882, Carl Rappold's family has ranched the Rocky Mountain Front just outside Glacier National Park, Montana. Seventy miles north, over the border on the edge of Waterton and Canada's International Peace Park, Charlie Russell's family spread dates back four generations. Both Rappold and Russell contend that in all those years, even when numbers of great brown bears were high, they never caused any problem. They credit mutual acceptance and cooperation by bear and human alike. Neither rancher shoots grizzlies nor do they discourage their presence with noisy hazing or calls for help to local wildlife officials. (Russell stopped carrying firearms in his teens.) By carefully studying the habits of grizzlies, the two ranching families learned how to go about ways of living in nonviolent coexistence with them.

Decades ago, when he was ranching in Alberta, Russell began leaving behind the bodies of cows and calves who had died. A cow carcass thus replaced the fallen moose or deer who, because of hunting pressures and the takeover of habitat by ranchers, were not as plentiful as in the past. A dead cow can satisfy a bear's desperate craving for a high-calorie meal as she transitions from lean post-hibernation to summer. With the ability to carry on as they did historically, bears have no need to prey on live bovine. Russell's method has finally caught on and local ranchers are following his lead with comparable success. Grizzlies have begun to rebound in numbers and size as a result of being able to live where and how they used to. There are reports that some bears now weigh up to a thousand pounds, whereas twenty years earlier they weighed, on average, only three hundred pounds.

The shift in attitude does not go unnoticed by grizzlies. Less wary and fearful of being hunted down, grizzlies are starting to relax. The Rappolds' expansive ranch shows just how casual and comfortable grizzly and human neighborliness can be. Every spring, grizzly mothers and cubs wander in to feed and play side-by-side with calves and cows, even going so far as eating from the same grain trough. Russell recalls a similar observation of bovine-bear compatibility while he was still ranching: "One really beautiful evening before sunset, I was sitting on my horse on a low hill overlooking a meadow where all the cows and their calves were lying down chewing their cud and relaxing, when I spied a big grizzly making his way towards us. He wove his way through the herd and not one cow made a move or showed any kind of alertness or fear. He finally got to the other side and disappeared into the aspen. I was amazed at what I just witnessed and decided to experiment. While still on my horse, I started to retrace the grizzly's path, but as I got to the edge of the herd, they began to get to their feet. Eventually, every last cow and calf got up and began to walk warily away from me. That sure was a lesson. The grizzly hadn't troubled them one iota, but my presence, even at a distance, was enough to disturb them."[12]

Nonetheless, the image of grizzlies as unyielding menaces persists. In the 1982 U.S. Fish and Wildlife Service's Grizzly Bear Recovery Plan, the celebrated Audubon naturalist and whooping crane activist Robert Porter Allen writes: "This is an animal that cannot compromise or adjust its way of life to ours. Could not by its very nature, could not even if we allowed it the opportunity, which we did not. For the grizzly bear, there is no freedom but that of unbounded space, no life except its own. Without meekness, without a sign of humility, it has refused to accept our idea of what the world should be like."[13]

Many take Allen's description of grizzly intransigence as a sign of the bears' belligerence. Most park managers and biologists decry laying down arms and maintain that grizzlies are about as reliable as a loaded gun. They insist that the species brings nothing but fatal trouble if allowed to live in undefended proximity to humans and furthermore, is aggressive even to its own kind.

Grizzly mothers can sometimes appear impatient and harsh with their offspring. Charlie Vandergaw, a longtime Alaska resident who has lived intimately with grizzly and black bears for over half his life, once witnessed a mother grizzly having a hard time convincing her three-year-old cub to start living on his own. Despite increasingly vigorous rebuffs from his mother, the young bear would not leave her side. Finally, she swatted him hard enough to knock him to the ground.[14] Although this incident may seem extreme, it is not rare to observe a gruff maternal response. Neither do adult grizzly fathers seem to make very good role models. They are known to stalk and kill cubs whose mothers fail to intervene in time.

On the surface, grizzly family life does not seem replete with what psychologists consider requisite conditions for healthy rearing—a warm, loving, and positive childhood environment. When a developmental environment fails to provide a child with appropriate nurturance, the result is often a suite of dysfunctional and asocial attitudes and behaviors. By this logic, rough grizzly parenting will produce rough young bears who grow up to be even rougher adults. This is exactly what wildlife managers and biologists assert: grizzlies are unpredictable, dangerous, and untrustworthy. But are they? Not when their behavior is examined using the combined lenses of Charlie Russell's experience, neuroscience, psychology, and wildlife biology.

In addition to his nearly lifelong experience with bears in Canada, Charlie Russell spent ten years rearing and reintroducing orphaned cubs from roadside zoos to the misty mountains of Russia's Kamchatka Peninsula.[15] It was not enough to know *about* bears to successfully ready the cubs for independent life. Raising a cub to think, feel, and act like a bear requires parenting that mirrors and models these qualities. His ten brown bear cubs (what grizzlies are called in Europe) acquired the necessary skill sets from Russell to find food, make a den, and live among fellow bears. He even taught the bouncing baby bears how to find and catch the precious salmon that make overwintering in Siberia possible. While some may argue that bear survival abilities are instinctual, the experiences of conservation reintroduction programs, such as the well-known program that reintroduced the California condor (*Gymnogyps californianus*), illustrate that many essential skills are learned, not inherited. A human parent-surrogate with insufficient understanding of the species' basic natural history or an inability to emulate what adult mammals and birds provide will fail to equip young animals with the necessary survival skills.[16] Like Daphne Sheldrick of the David Sheldrick Wildlife Trust in Kenya, who has rescued hundreds of orphaned infant elephants, Russell was able to understand an animal mother's mind enough to know what and how to teach her young.

The idea of a bear psyche is really not new, nor are the workings of that psyche much different from our own. Core brain structures and mechanisms involved in early development are evolutionarily conserved across species.[17] Importantly, too, cortical and limbic structures related to maternal and other affiliative behaviors and psychological states are shared among all mammals.[18] This means that what we know about human development can be extended to bears—the same principles apply.

That is not to say that differences between species do not exist. There are reasons that bears turn out to be bears, pumas to be pumas, and humans to be humans. But as the Notre Dame neuropsychologist Darcia Narvaez underscores, differences in psychological outcomes derive less from inheritance than from epigenetics and early development: in other words, "biology is not fixed at birth but is shaped by experience." It is the contrast between grizzly parenting values and those of modern humans that leads to perceptual and behavioral differences among their youngsters. The early experience of infancy and childhood, "when caregiving shapes brain and body, engraves neurobiological systems and functions that underlie the ethical dispositions guiding one's social life."[19] Grizzly cubs exercise grizzly rules because they are taught to do so, and so it goes with humans.

Developmental experiences dating from inception through childhood are strong predictors of infant mental and emotional development, which in turn strongly influence who we grow up to be. Although neuroimaging reveals that brain maturation continues into adulthood, the most rapid neurobiological change takes place during the first months and years of life (the precise time frame depends on the lifespan of the species). Key players in life's unfolding are parents and families.

Human parenting in the West went largely unexamined until the nineteenth century. Child care was culturally embedded in and shaped by prevailing social and economic circumstances. How parents raised their children was the business of the family and the family alone. This all changed with the disruptive effects of colonization and urbanization that overturned millennia-old social traditions.[20] Outside ruling entities—employers, governments, and school authorities—suddenly appropriated power over child care that had historically been vested in the tribe or family head.

By the mid-to-late 1800s, the fallout from industrial serfdom and the destruction of indigenous cultures had become increasingly evident. Children were part of this social flotsam. Successive wars shredded social cohesion, leaving millions of children psychologically wounded and in want. As debates on child labor and protective laws and rights heated up, educators and activists started to scrutinize methods of upbringing. At the same time, psychologists such as Pierre Janet, Sigmund Freud, and C. G. Jung brought scientific attention to the link between childhood and adult mental and emotional wellbeing. Many of these ideas were distilled into British psychologist John Bowlby's conceptual framework describing systematic patterns of relational development, or what he termed *attachment theory.*[21]

Bowlby's theory took form in the aftermath of World War II, during his work with orphaned, abused, and abandoned children. After extensive study, he found an emerging pattern that related child psychological states and behaviors to their rearing experiences. Those who had been provided with parental care that was responsive, attentive, and loving were able to cope with change and stressful situations. Although they might naturally respond with distress and fear, the children could generally regain a sense of equanimity in the wake of deprivation or violence. Their engagement with others mirrored the empathetic and centered relationships that they had experienced early on. Bowlby defined this formative, positive relationship a *secure* attachment. Just as he foresaw, neuroscientists were later able to relate what psychologists had observed externally to what was happening on the inside.

Allan Schore is a clinical psychologist on the faculty of UCLA's Department of Psychiatry and Biobehavioral Sciences, David Geffen School of Medicine, and Center for Culture, Brain, and Development. He is also the author of three path-breaking books on the neurobiological mechanisms of attachment and trauma. Schore pioneered the merger of attachment theory with neurobiology by mapping Bowlby's descriptions of caregiver-baby social interactions to nascent brain structures and processes.[22]

The everyday social and emotional exchanges involved in bringing up baby don't stop at the skin: they penetrate inside and shape the circuitry of an infant's highly plastic, receptive developing brain. Nuzzling, cooing, laughing, nursing, and eye-to-eye contact all tune the brain's neuroendocrine pathways, which in turn govern emotions and stress regulation.[23] By engaging the brain's cortical system, attachment processes regulate more instinctive behavior and emotions. Or as Schore explains, this mammalian control system "can now be identified as the orbitofrontal system and its cortical and

A Grizzly Bear secure attachment. Photo credit: C & C Bear Imagery.

subcortical connections [where the] 'senior executive of the emotional brain' acts as a regulatory system, and is expanded in the right hemisphere, which is dominant in human infancy and centrally involved in inhibitory control."[24] In substance, attachment involves interactive regulatory processes that synchronize the brains and minds of two individuals, the infant and the caregiver. Mother-baby exchanges are inscribed in the infant's cortico-limbic system of cognitive checks and emotional balances.[25]

Such relational molding can be healthful or unhealthful, depending on the quality of interactions. Not all childhoods are positive, nor do all parent-infant exchanges emulate a secure attachment. When the relational molding process is negative, a child experiences what Bowlby and his colleague Mary Ainsworth have termed an *insecure attachment.* Distressing, unstable emotions and thoughts in children are correlated with neglectful and abusive parenting. Even children who do not experience abuse in the technical sense, but whose parents are unresponsive or inconsistently responsive, may develop an insecure attachment. In a series of now famous studies, the "still face" experiments, developmental psychologist Edward Tronick vividly demonstrated how an insecure attachment is generated. Videotaped sessions with a mother and her baby or toddler depict how the slightest, almost imperceptible, change in a mother's mood and facial expression is immediately telegraphed to and detected by the baby.[26]

Relational discord and its resultant cortico-limbic imbalance frequently lead to emotional distress and dysregulation. The effects of early relational trauma typically produce an enduring deficit throughout an individual's lifetime, which makes it difficult for that person to cope with stressful or emotional experiences.[27] Bowlby observed that children who were neglected or treated violently expressed aggression, hypervigilance, an inability to regulate feelings, sociopathy, and anxiety, particularly in their social interactions.[28]

Statistically, children who experience trauma or emotional neglect are at greater risk later in life for psychological disorders, compromised immune systems, and a suite of diseases. Physiological and emotional vulnerability may even develop prior to birth. Maternal stress experienced during gestation—trauma, hunger, chronic or extreme fear—affects the neurobiology of the developing embryo via stress hormones. Stress transmission also occurs socially. Stressed or anxious mothers often exhibit decreased positive, nurturing responses to their offspring, which in turn increase stress reactivity in those offspring.[29] After birth, elevated levels of anxiety and fearfulness by the baby experienced in utero may express behaviorally. What we experience translates into how we act.

Psychologists recognize three general types of insecure attachment styles: avoidant, ambivalent, and disorganized. All involve some sort of "undercare" that renders the sensitive infant brain acutely reactive to stress and effects an overall shift from an innate attraction to nature to alienation from it.[30] Avoidant behavior is associated with an *avoidant insecure* or indifferent-attachment parenting style. These children tend to be self-reliant to the point of being detached and unfeeling toward others. In adulthood, they may have difficulty sharing intimate feelings and committing themselves socially or romantically. Children who experience inconsistent and conflictive care from their mother or parents often exhibit a pattern of *ambivalent insecure* attachment. Similar to their caregivers, they alternate between possessive closeness and rejection. Their personalities and behavior are characterized by inappropriate dependency and hyper-anxiety. As adults, they may show correspondingly high levels of distress and ambivalence toward others. *Disorganized insecure* attachment develops when children are abused. A child's natural need for care and protection from a parent is left unmet and conflicts with the parent's threatening actions and the child's fear. Later in life, this contradiction translates to anxiety; inconsistent, erratic behavior; and confusion around relationships.[31]

The fact that early attachment experiences are wired into the brain accounts for why many find it difficult to change patterns of thinking, behaving, and feeling. How we learned to respond to the environment as infants accompanies us throughout life.[32] Because the natal environment informs a child's reality, and that reality tends to perpetuate in other, adult relationships, it may take years before the child-now-grown-up becomes aware of the source and influence of these dynamics. Neuropsychological resistance to change is not always detrimental, however. There are benefits in terms of both survival and the intergenerational transmission and perpetuation of healthful states.

Psychiatrist Henry Krystal, a Holocaust survivor, worked with concentration camp victims throughout his lifetime to support their recovery. His longitudinal studies of survivors revealed that those who managed to live and afterward craft some semblance of well-being tended to have had secure attachment experiences early in life. The inner sense of well-being and security provided by a loving family functioned as a kind of vaccine that supported the resilience needed to withstand the horrors of the camp: "even when confronted with trauma, the individual is able to access the supportive caregivers and values that have been internalized unconsciously in childhood

Bisquit and Charlie relaxing in trust. Photo credit: Charlie Russell and Maureen Enns.

and [to believe] that 'the restoration of one's capacity of love'" is possible.[33] Notably, a secure attachment was determined less by familial structure than it was relational quality. *How* was more important than *who.*[34]

The critical, vital nature of attachment for physical, as well as mental, survival is vividly illustrated by former SeaWorld trainer John Hargrove's account of a motherless, captive-held orca who had not been taught to breathe in air when beached. Because the young orca was "not fully trained on the medical pool," the trainer and staff had "to get this whale to breathe, [and so] we had to lower the med pool floor, but then the animal would immediately start thrashing and smash into walls. Cuts start to appear on the body, the whale is bleeding and making vocalizations of a panicked whale. These vocalizations are similar to when a really dominant whale is attacking a subdominant animal. It's that type of vocalization but just off the Richter scale because it knows it is dying. It's a horrible sound, short, sharp vocals. It's thrashing around with eyes wide open and you can just see the panic."[35]

Attachment provides the implicit and explicit learning needed to help infant orcas, bears, and infants of other species navigate the ups and downs of their dynamic environment.

As attachment theory took root and expanded to broader contexts, mental health workers turned a critical eye toward modern parenting, which, when examined from a cross-cultural standpoint, stood in stark contrast to both human and animal societies.

Statistically, modern models and methods of child-rearing are a blip on the scale of human history.[36] The anthropological record shows that small-band gatherer-hunter societies "encompassed 99 percent of human genus existence," which argues that their respective values, culture, and behaviors, not those of the modern West, form the logical "baseline range for human society and human development."[37] On the whole, human tribal societies reflect collective, prosocial values—empathy and an ethic of care for others—which foster a culture of cooperation and companionship. Similar to elephant society, gatherer childcare draws on a web of parents, aunts, siblings, and nonhumans.[38] From this vantage point, nineteenth- and twentieth-century child care, once touted as the gold standard of rearing patterns, appears to developmental psychologists such as Darcia Narvaez to be responsible for much of the social dis-ease afflicting current societies.[39]

Starting in the 1700s, or even earlier, the ideal human to which Western child-rearing aspired was a rational, self-contained, and autonomous individual "locked within the privacy of a body," "standing against" and competing for the "rewards of success" with "an aggregate of other such individuals" in the society. Events of the nineteenth and twentieth centuries—including widespread industrialization, cultural homogenization, and the violent chaos of successive wars—reduced the once complex molecule of social function to the nuclear family, that is, to the bare bones of a mother, father, and siblings. The seeds of present-day parenting, then, were planted well before this period, during the Enlightenment. The result? In comparison to both animal societies and human gatherer-hunter societies, Westerners generally have anonymous, "brittle, contingent, and transient" relationships that lack "direct, intersubjective involvement."[40] Further, when human and animal societies are compared, differences among species seem less significant than variations across human cultures. In particular, gatherer-hunter societies seem to have much more in common with elephant and other animal kinship groupings than they do with their modern, Western conspecifics.[41]

While psychologists were delving into the minds of children and parents, ethologists were constructing their own views of development. It is not as if Bowlby was unaware of this parallel course. His treatment of attachment theory and his seminal trilogy are grounded in animal behavior and evolutionary theory and laced with examples from the animal world. At that time, during psychology's teen years, there was ample interdisciplinary exchange between ethologists and psychologists, which is one reason that attachment theory has found its way across various academic fields and species. The salient difference in how the two fields have approached the study of development is that psychologists largely focus on interpersonal relationships, while ethologists frame rearing in terms of the developmental context—the environmental microcosm into which infants, cubs, fry, and chicks are born.[42]

In some ways, students of animal behavior have been much more aware of the variability in and quality of rearing patterns. As zoologist Robert Hinde noted, "There is not a best mothering (or attachment) style, for different styles are better for different circumstances, and natural selection would act to favour individuals with a range of potential styles from which they select appropriately."[43] Similar to the beaks of Darwin's famous finches that evolved to match resources and needs of their surroundings, behavior has been conventionally viewed as adaptive, a passive product of nature's selective forces. Species-specific rearing styles maintain only if they remain appropriate to the conditions in which a given animal must survive and successfully reproduce.

Psychologist Jay Belsky drew on this ethological-evolutionary chain of logic when he proposed that normative attachment may not always be a secure attachment. Prosocial behaviors may be contraindicated in favor of insecure attachments in communities dealing with the aftermath of genocide, war, or high levels of violence.[44] Chronically high-stress environments create huge demands that require behavioral adaptations and neuroendocrine, metabolic, and cardiovascular responses that are able to compensate for the external stressors of constant danger and resource shortages. Maternal stress that expresses as indifference and aloofness transmits to the next generation as self-reliance, aggression, hypervigilance, anxiety, and insularity—psychological states that ready an individual for situations where there exist ever-present danger, untrustworthy people, and a high probability of assault.[45] In other words, the argument goes, creating infant development patterns that anticipate the larger environment in which they are expected to successfully navigate—even if it is an insecure attachment—is evolutionarily adaptive.[46] For a violence-defined, sociopathic environment, insecure attachment styles are fitness-enhancing strategies.

Obviously, the idea that parental abuse and indifference could be evolutionarily optimal strategies clashes with the image of what is considered a healthful, ethical, prosocial environment.[47] It also runs contrary to both the anthropological record of human social behavior and developmental models that take into consideration the important effects of epigenetics and sociality. The Developmental Systems and Evolved Developmental Niche theories integrate nature and nurture and are informed by consideration of the fundamental social function of the brain.[48] From this view, an insecure attachment undermines survival because it interferes with the development of socioemotional intelligence, considered one of the most significant evolutionary characteristics of humans and other social animals.[49] Furthermore, as Darcia Narvaez points out, functional adaptation (the ability to conform to one's life situation the best that one can) is not the same as genetic fitness (the ability to outcompete a rival over several generations). The germline of an abused child will not outcompete that of a well-raised child because of impairments at many levels of functioning.[50] Socially astute individuals were evolutionarily successful (that is, in the currency of evolution, they and their progeny came to comprise 99 percent of the *Homo sapiens* record), for the very reason that they were able to get along with each other. Prosocial ways led to prosocial brains, prosocial brains to prosocial ways, and so on down the genealogical tree.[51]

Psychologist Katharina Rutschky calls the trend of insecure-attachment-style parenting, and its attendant asocial and anti-social outcomes that have dominated the West, a "poisonous pedagogy" (*Schwarze Pädagogik*). Psychoanalyst Alice Miller quotes copious references from as early as the 1500s that promote a normative attachment pattern based on the violent breaking of children "for their own good." For example, a passage from *A Godly Form of Household Government,* a leading "conduct book" of the time co-authored by Puritan clergymen John Dod and Robert Cleaver, reads: "The gentle rod of the mother is a very soft and gentle thing; it will break neither bone nore skin; yet by the blessing of God with it, and upon the wise application of it, it would break the bond that bindeth up corruption in the heart . . . Withhold not correction from the child, for if thou beatest him with the rod he shall not die. Thou shalt beat him with the rod and deliver his soul from hell." Such "earliest experiences unfailingly affect society as a whole" and many believe that our present-day epidemics of "psychoses, drug addiction, and criminality" are "encoded expressions of these experiences."[52]

As framed by historical and anthropological data, abusive, neglectful parenting was perpetuated in the West not as an evolutionarily successful adaptation, but because of its psychological and cultural intransigence. "Probably the majority of us belong to the category of 'decent people' who were once beaten," Miller asserts, "since such treatment of children was a matter of course in past generations. Be that as it may, to some degree we can all be numbered among the survivors of 'poisonous pedagogy,' yet it would be just as false to deduce from this fact of survival that our upbringing caused us no harm as it would be to maintain that a limited nuclear war would be harmless because a part of humanity would still be alive when it was over."[53] It is certain that our ancestors sometimes encountered less than favorable conditions, yet somehow, as the record shows, our forebears continued to make the choice toward prosocial values—until, it seems, the recent past.

What exactly lay behind the choice by Rutschky's psychic poisoners to take the childrearing paths less traveled is mysterious and debatable. But the questions at hand are: Do these perspectives relate to grizzlies' development? Do grizzly cubhood experiences account for the conventional scientific characterizations of bears as unpredictable, aggressive, and ill-suited to prolonged exposure to humans? It is not farfetched to imagine that if popular depictions of grizzlies were passed across John Bowlby's desk, he might diagnose the species' attachment style as insecure. He might even go so far as to predict that grizzly youths are likely to exhibit aggression, belligerence, poor emotional regulation, and other symptoms of psychological distress. But when details of bears' natural history are examined more closely and accurately, we find that grizzlies are masters in the art of secure attachment parenting—and it all starts in the cozy quarters of the natal den.

Unlike many other species, grizzlies begin their lives during maternal hibernation, and motherhood is well under way even before the bear eggs begin to develop in her womb. When a female bear becomes pregnant, which is usually sometime between late spring and summer, fertilized egg implantation is not immediate. Before infant care begins, there is precious food to gather and calories to be stored to provide enough rich, plentiful milk for her babies, and enough fat to overwinter and keep her fit for the weeks after her spring emergence.

The time of denning is carefully chosen: the mother-to-be bear waits for just the right moment. Finally, after eking out every last berry and salmon carcass, she meticulously prepares a den that will afford appropriate space and protection for herself and for the baby bears born during the winter retreat. When the cold winds of late fall begin to blow, warning of winter, the mother bear, fattened by hard-earned fruits of summer, searches for a place and digs a space that, much like Goldilocks's criteria, is not too small and not too big. Typically, incipient mothers begin to den sooner than other bears. Young bears and other single females den later, followed by male bears, who seem to hold out to the last before hibernating. When winter melts into spring, the order of emergence is reversed, with male bears venturing out first.

Bear dens are a marvelous invention. They vary in substrate and design by species, geography, and climate. For example, while black bears tend to dig dens under windfalls and hollow trees or in caves, grizzlies seem to prefer the base of a large tree. Choice usually depends on optimizing protection from the elements and intruders and is influenced by slope angle, prevailing wind direction, and other related factors including, as Charlie Russell has observed, aesthetics. Although views may be less important for dens than for summer day nests, Russell found that when he sat in the empty nests to get a "bear's eye view," all were situated and oriented in ways that were aesthetically pleasing.

Taking regional and local variation into account, the average architecture of a grizzly den consists of an entrance, a short tunnel, and a sleeping chamber. In Siberia, the sleeping chamber is often built above the entrance tunnel to create a bench. There are several obvious advantages to this design. It keeps out runoff, provides a high-ground defensive advantage should a predator venture in, and creates a place where cold air may settle so that any warm air stays right where it needs to be—with the mother and cubs.

There are other bear den styles. A tight bunch of saplings, for instance, can provide the underlying structure of a snow-covered roof. Charlie Russell once came across the den of two cubs he had raised in Kamchatka. It was the young bears' first hibernation alone, and they had crafted a very neat winter den. Beneath a bower of pine saplings, the den's earthy floor had been dug up a little and padded with branches and grass. When snow drifted in, it bent the saplings into an insulating arch that created a snug refuge. This kind of hibernation dwelling does not work in places where dangerously low temperatures can set in before snow, such as the Front Range of western Canada, where grizzlies historically have proliferated. Bears there have other tricks up their hairy sleeves, such as waiting to den up until a severe snowstorm hits so that their tracks are hidden from potential predators. The bear is a clever fellow.

Natal dens have other amenities. The mother bear will usually drag in spruce branches, ferns, or similar nearby material to create bedding that makes the winter nest better insulated and more comfortable. Denning is no guarantee of survival, and in the harsh surrounds of Siberia and subarctic North America, every little bit counts. Woodsmen describe coming across either the carcass of a bear on the snow or an already-expired bear in a den. Death in winter can be caused by many reasons, including insufficient calories stored for hibernation, a bear's age and health, and inadequate climatic and den conditions. Whether they were ill-prepared, too inexperienced to successfully confront the winter conditions, or poor craftsmen, some bears are forced to abandon their lair during winter and dig another if the den fails to do its job properly (for instance, if water starts to seep in).

In hibernation, bear metabolic rates decrease by half, and breathing rates can slow to not much more than once a minute. Bears live off the fat that they have accumulated during the warm months, and any waste is recycled.

Buck, formerly orphaned, swimming with Charlie. Photo credit: Charlie Russell and Irina Kruglyakova.

But unlike other hibernators such as ground squirrels, bear body temperatures remain near normal, and they usually do not wake periodically. For these reasons, a well-insulated den supports the hibernation process by ensuring that very little precious heat is lost. When bears emerge, they may have lost up to a third of the weight they had before denning.[54]

Fertilized egg implantation usually takes place in November or December, and the cubs are born six to eight weeks later.[55] A mother bear may give birth to as many as four tiny—only a few inches long—hairless, blind, and extremely vulnerable cubs. Their survival is totally dependent on their mother and the den she has created. Cub brain and emotional bonding begin in the glow of a tender, nursing mother and siblings. In this womblike enclosure, the young bear is tuned mentally, emotionally, and physiologically for several months to nothing and nobody but grizzly language, values, and ways.

Every smell, movement, taste, sound, and when their eyes eventually open, sight, is grizzly informed in the cozy, all-encompassing world of the natal den.

Throughout hibernation, infant bears nurse on their mother's rich milk and remain in this hermetically sealed world until they emerge in late spring to early summer. The spring sun first kisses the faces of the cubs when they are between four and five months old. At this point, the cubs leave their bubble of love and venture outside. Emergence from the natal den into the big world marks a second developmental phase. Life in the blossoming woods, streams, and meadows is pulsing with animals busy looking for food or readying for mates and families. But although there may be some interfamilial interactions, particularly at river's edge when bears fish almost shoulder-to-shoulder, the new family forms a fairly autonomous constellation. Much like a stable ship with tethered lifeboats, mother and cubs make their way as they transition from den to wide-open world.

Bringing up grizzly cubs is not easy, especially for first-time mothers who are still getting the lay of the land. A mother must toggle between keeping an eye on the vulnerable cubs and searching for food. She is responsible for providing her young with food, care, affection, and formidable guardianship against larger, stronger male bears and other potential predators. In this often tricky social and ecological terrain, the mother bear has to maintain a stable psychological, emotional, and physical domain within which her cubs' young minds can develop. Washington State University bear researcher Charles Robbins asserts that mother bears and mother humans are very similar in how they care for their young and their concerns: "They worry about them like a human parent does."[56]

From that unwavering base, the cubs can focus on other life skills such as foraging, building up muscles and coordination, and learning the ins and outs of grizzly culture and living in the wild.[57] Cubs engage daily in Bowlby's cycle of attachment, separation, and detachment—a developmental dance alternating between union and individuation that leads to well-coordinated, balanced bear relationships—or from an inside, neuropsychological perspective, a well-coordinated, balanced cortico-subcortical relationship.

An infant uses three basic strategies to keep close to her mother: positive gestures and vocalizations (playing and inviting), negative gestures and vocalizations (cries), and actions (running after). All of these interactions shape social and neurobiological expectations in anticipation of eventual weaning and adulthood. Secure attachment parenting means that the mother will respond appropriately to all three of these efforts on the part of the cub. The quality, timing, and nature of these interactions are carefully choreographed. When the mother does not respond, the cub often expresses despair and despondence and a cessation of movement, in what some suggest is an evolutionary adaptation that decreases interest by predators. If the mother continues to be unresponsive, the cub will begin to detach emotionally from the mother. Such a detachment from the attachment figure effected at the right moment in the developmental arc opens the door to new relational bonds—an openness not only to siblings but also to membership in the broader socioecological environment.

As summer ripens, so does the menu. Grizzlies shift to a more calorie-rich diet, which includes pine nuts and salmon. The fish are also on a pressing schedule. They have traveled hundreds of miles from the ocean to streams that cascade down rugged mountains and into the lakes they feed. Whereas the bear cubs are at the dawn of their life cycle, the salmon are in their sunset days, struggling back to natal waters where they will spawn and die. Like two waveforms of opposite signs, the life arcs of grizzlies and coho come together in a perfectly orchestrated ritual.[58] As the colors of the season turn to reds and yellows, the family begins on its path to hibernation. While in the den, they will replay the experiences of the previous year by spending five months together in another natal bubble.

During subsequent summers, and a couple months following hibernation, a mother bear may leave her cubs for a week or so to mate again. This is not a hard and fast rule, however. Some grizzly mothers return to their cubs and even den with them a fourth time while they are pregnant, only to oust them when the new young are born so there is enough room. Finally, when weaning and complete separation do occur at the age of three years, the relational tear is notably emotional. Cubs struggle to stay with their mother, seeking to engage her affectionately or crying out in distress. Eventually, they are met with "tough love," or increasingly stern responses that sometimes, as discussed earlier, reach the point of physical violence. In the aftermath, the "teen" cubs, though as large as young bears, nonetheless seem despondent. Not infrequently, siblings remain together, and when winter threatens, den together.

Separation from the mother readies the young bear for his or her adult life. With the exception of short-term close interactions such as mating and hanging out together at food-rich salmon streams and berry patches, females and males tend to live on their own. Males will battle during mating season and some will stalk cubs. Significantly, not all males prey on cubs; rather, it seems that there are certain males who "turn" cannibalistic. Some assert that a male will kill cubs so that the mother will go into estrus and he can then mate with her. Charlie Russell questions this explanation: "Not all male bears go after cubs. They have opportunity, but don't pursue it. The rare few that do hunt cubs, go at it tenaciously. There was this big male bear who showed up one day. I called him the unimaginative name 'Killer.' He began scoping out our 50-meter-square electric fenced pen that I had built to protect the cubs when they were small and still very dependent. The bear would check out the pen from all sorts of angles trying to figure out how to get in and grab a cub. He got shocked once so he was respectful of the fence but he would not let up . . . I never felt personally at risk. He never went near the cabin. He was only intent on getting my cubs."[59] Charlie adds wryly, "Killer was in no way interested in me, not as food, and certainly not as a mate."[60] It is possible that cannibalistic tendencies are triggered as a result of one or more factors: a compromised upbringing, poor health, or end-of-season food shortages. A fat cub provides a handy source of calories for a needy bear.

Subsequently, despite grizzly bears' reputation as frightening loners, when details of their natural history are examined more closely, a much more endearing picture emerges. From implantation through their third year, grizzly young live in the protective care of their mother and siblings. During this critical period of mammalian social and emotional development, mother and cub not only spend up to thirty-six months together; they also spend almost half this time snuggling inside a winter den, engaged in prosocial bear brain-to-brain synchrony. The close quarters form a developmental cocoon that encases mother and cubs and guarantees deep, sustained, and highly specific neuropsychological tuning. From the perspective of evolutionary psychology, then, cub attachment experiences fall within species-specific psychophysiological expectations. From the perspectives of Developmental System and Evolved Developmental Niche theories, grizzly prosocial development naturally leads to prosocial adulthood, behaviors of which are consistent with statistics and observations that grizzlies are not hair-trigger aggressive.

By Bowlby's criteria, female grizzlies qualify for the Mother of the Year award. Further, symptoms correlated with childhood relational trauma prevalent in the human communities described by Belsky are absent among grizzlies. Attachment disorders emerge only when the early social environment does not provide for the child, and it is clear that grizzly mothering does. After weaning and separation, the young grizzly is fully equipped to survive and thrive foodwise and socially. Unlike human children in violence-torn communities, the relational tear experienced by young bears is apparently not traumatic; rather it is context-appropriate and congruent with a successful grizzly lifestyle trajectory. Generation after generation of bears repeat the same mothering style of affection and protection that begins in the nurturing micro-culture of the bonding den. Young bears are inoculated with the kinds of support that Henry Krystal considered vital—parental love and security.

Charlie Russell, when raising orphaned cubs in Russia, emulated mother bear behavior that he had observed. His intuitive cultivation of a secure bear attachment was in every way the definition of prosocial—a centered, nonreactive, positive way of being that involves empathy for bears as well as humans. In essence, Charlie internalized the model of a mother bear, thereby molding himself, as closely as possible, to become an appropriate partner in the socio-affective tuning of young bears. His modeling shaped cub brains and minds in ways fitting to the environment in which the child-turned-adult bear is required to operate successfully—at least before illegal hunters began slaughtering them.

There are those who may argue with the characterization of grizzlies as social, let alone prosocial. While the term is generally used when describing human behavior, prosocial extends to other animal species. Researchers have documented altruism and concern for conspecifics among species as diverse

Charlie teaches Bisquit where to find salmon.
Photo credit: Charlie Russell and Maureen Enns.

as walruses, ravens, African buffalo, elephants, lions, and even army ants. Prosociality does not presuppose any particular social geometry. Hence, even though grizzly sociality may not be as tightly orchestrated spatially as African lions, for example, who live in a pride, bears do run on a set of rules that comprise a kind of social compact. Studies on the relationship between food abundance and social patterns found that black bears behave "in accordance with kinship theory within a social order governed mainly by the distribution and abundance of food." The typically more solitary bears even form social hierarchies in concentrated areas of food such as garbage dumps.[61]

In addition, a lot more communication among bears takes place than most humans can perceive. Unlike their canid and feline cousins, grizzlies do not use "weapons threat" gestures such as lip curling to communicate, nor do they sport long eloquent tails. Instead, bears utilize distinct trails, "highways" along which they travel, and specific trees where they rub their backs and leave word for future, passing bears. Bear country is riddled with intricate bear trails and marker trees that serve to convey information in the absence of bear-to-bear physical proximity. These communiqués create an invisible web of messages that brings coherence to a physically dispersed community. Bears' psychology is thus uniquely suited to their habitat.

We might then envision bear collective life as a dynamic network of relationships that is shaped by resources and terrain, much of which, similar to those used by other conventionally regarded "cryptic" or "covert" species, is hidden from human eyes. As we will see in successive chapters, biologists are discovering previously unsuspected psychological depths in species that traditionally have been classified as subsocial, parasocial, asocial, or other deviations from true sociality, or eusociality.

There are also indications that inter-bear relationships are a bit more complex and that what are usually categorized as the simple necessary basics—mating, parenting, predation, and so on—may have many dimensions. During the construction of his cabin in the Alaskan wilderness, beginning in 1975, Charlie Vandergaw noticed that more and more bears—both grizzlies and black—were showing up and, as he puts it, "just hanging out" at what came to be called "Bear Haven."

> Bear Haven was not my original intention. It just evolved on its own. As the years passed, I realized that it afforded the bears a place where they could meet, mingle, and express themselves in ways that they can't elsewhere. Bear Haven became a kind of crossroads' oasis in a matrix of otherwise hostile, tense hunting ground. Early on, I once saw a large black bear and large female grizzly in the yard. The grizzly was entirely at peace and the black bear too. I don't know whether they knew each other—both were strangers to me. That happened a lot as time went on. Mothers of both species brought their babies to show and share with me, but also with each other. At one time, there were three generations of grizzlies. They didn't bother each other and they never bothered me. The mothers taught their cubs the rules that I taught them and we were all able to get along. My part was that I em-

> braced, not displaced, the bear neighbors. It was a place of mutual respect and peace. That is why it worked. Instead of telling the bears what to do, I listened.[62]

Some may protest that the provision of food supplements at Bear Haven created an unnatural environment and hence the observed bear behaviors were not natural either. But the food provision does not change the fact that the bears were socializing among themselves, much like they would when together at salmon-laden streams where they belly-up to the bar side by side, with the abundance of food leading to decreased social tensions.

On the whole, then, grizzly behavior *is* predictable. Within grizzly society, rules of conduct are followed precisely. Conflict in fish-dense rivers and intraspecific fatalities involving competing males or mothers fighting off predatory males are rare. Despite their formidable power, grizzlies do not exercise it often against each other or against other species, humans included. The nonaggressive nature of grizzlies is reflected in the records kept by those who have become well-schooled in the language of the wilderness. In their view, grizzlies are dangerous only when their rules are not being followed or when they are under attack.

Echoing Muir, Mills, Rogers, Van Tigham, and many others, Russell maintains that dangerous bears are created by people—in their minds and by their actions: "Basically, I am not fearful of bears. This does not mean that I don't get scared on occasion, but the main feeling I have is being relaxed and open—and I trust them because I guess I trust myself and what I can read in bears and the landscape. It is critical to not expose yourself foolishly and you must always remember that every bear carries a unique perspective and experience. The three bears involved in the three fatalities with which I am familiar were distinct in terms of their past, body and mental condition, and interactions with humans. This also includes specifics of what each human did or did not do and their individual attitudes. When I looked at the facts of these incidents in detail, the single most common factor among them was the difference between the men and the bears."[63]

Russell's point is that, as we saw in the case of sharks, individual idiosyncrasies often play a more important role than species generalities, even though wildlife scientists tend to rely on these stereotypes. Albert Memmi observed this same view in North Africa. French colonizers imposed racist agendas by redefining colonized human and nonhuman indigenous communities as a homogenized single type. This "mark of the plural," the refusal to see the individual apart from the stereotypic and racist gaze, was a central ingredient in the colonial mindset. Similar to North American grizzlies, individual identities and the agency of the African colonized were erased and rendered into background—making them an anonymous substrate whose sole purpose was to serve the colonizing French, or in the case of wildlife, to serve modern humans who have colonized grizzly land and condemned bears through mythic caricature.[64]

Historical testimony documenting that Kamchatkan women felt at ease with grizzlies, even in the presence of food and children, shows that human habituation is also not the problem. This is reified by Lynn Rogers's decades-long work at the Wildlife Research Institute, in which he has investigated the effects of feeding black bears and discovered that bears are more than ready to make peace with humans. In parallel, recall as soon as grizzlies in the Front Range, Canada, realized that they are no longer targets of human wrath and guns, they readily reoccupied human-inhabited areas and proceeded to go about their business. So, as we circle back, we find that what Muir, Miller, and Russell have observed concurs with what psychology and the neurosciences predict. Despite the myths that dog their reputation, grizzlies are indiscriminately prosocial and their supposed ferociousness is only because, as Enos Mills puts it, they "object to being killed."

That said, however, it would be surprising indeed if grizzly brains have not been affected by humanity's ceaseless persecution. Grizzlies and other wildlife have been under siege for hundreds of years enduring chronic stress and trauma and the renting of their socio-ecological fabric. Loss of habitat; dramatic declines in numbers from poaching, hunting, culls of older bears, and disease; translocation; and the ever-present threat of violent human intrusion have transformed grizzly habitats into landscapes very much like human war zones. These stressors are comparable to what in human terms, and from neuroscience's species-leveling paradigm, would be called murder, genocide, ritual killings, deportation, appropriation of homeland, and ter-

The Brown Bear who sat every night watching the sun set, Kamchatka. Photo credit: Charlie Russell.

rorism. The experiences of the grizzly are not much different than those of their human tribal counterparts, American Indians, who have been subjected to what Maria Yellow Horse Braveheart terms *historical trauma*—that is, the "cumulative emotional and psychological wounding over the lifespan and across generations, emanating from massive group trauma experiences."[65] Tragically, similar to love, its antithesis, the experience of violence, passes on intergenerationally.

By any measure, the foundations of traditional grizzly society have been severely challenged. Given bear mortality rates from hunting and other human causes, it is no exaggeration to write that every bear has experienced directly or indirectly one or more traumatic events that would qualify him or her as a candidate for the diagnosis of PTSD. Charlie Russell now carries bear spray. "I can't use the same rules with the bears here like I used to because they have been beaten up so badly by humans—guns, hazing, noise, translocations and all kinds of bullying. Vitaly Nikolayenko, the Russian bear expert, told me that bears pass down their fears to their cubs for several generations to keep them from having to learn the fear first-hand in the face of actual danger."[66] As a result of human assaults and public policy based on intolerance and fear, bear behavior may be straying from its past, predictable rules of engagement. The seedbed of wildlife consciousness may be deep and rich, but the grizzly mind of today is not the grizzly mind of the past.

3

Orcas: Sense of Self and Moral Evolution

Who are you? Take a minute and write down what comes to mind. Maybe you'll say your name or what you do for a living—plumber, teacher, student, driver. You might add your gender, ethnicity, or nationality. Next, think about someone asking you the same question somewhere else—at a party, walking in your neighborhood, at a pickup basketball game after work,

or another situation. You might then answer, "Barry's sister." Or "the neighbor two doors down." Or "the blue team guard."

Although there may have been a momentary existential pause (like "Good question, who *am* I?"), your responses indicate self-awareness. That you relate to an "I" means you are self-aware. According to scientists, this is a significant achievement in the animal kingdom. Ever since the 1970s, when psychologist Gordon Gallup came up with the idea of using mirrors to test for self-awareness, there has been a flurry of investigation into whether animals other than humans possess this capacity.

Equipped with an assemblage of mirrors, paints, and protocols, researchers have dashed from species to species trying to see who fits the glass slipper of self-awareness. If the candidate animal passes the mirror test by indicating that she recognizes herself in the reflection, then she is given membership in the elite group. So far, many species have qualified—chimpanzees, elephants, dogs, dolphins, and the spectacular black and white whale, *Orcinus orca.*[1]

Orcas are the largest animals of the dolphin family, which belongs to the broader classification of whales, Cetacea. This infraorder then splits into two groups of whales—baleen and toothed. Baleen whales, also called mysticetes, are filter feeders who gather their food (crustaceans and plankton) by sifting rich ocean waters using keratin bristles. Thousands upon thousands of right, blue, humpback, and gray whales were slaughtered for oil and for this fibrous dentition that was used in corsets, collar stays, and other clothing.

Toothed whales, Odontoceti, include the familiar bottlenose dolphin (genus *Tursiops*), harbor porpoise (*Phocoena phocoena*), pilot (genus *Globicephala*) and sperm whales (*Physeter macrocephalus*), and orcas. Dubbed "killer whales" because they hunt and kill other whales, orcas carry a full set of four-inch-long, pointed teeth on their upper and lower jaws. Orcas are supreme apex predators capable of stalking, attacking, and eating their fish counterpart, the white shark, in just a few bites. As we saw in the case of the hapless white shark off the Farallon Islands, when a young orca, CA6, flipped him on his back with the agility of an Aikido master, orcas' skills combine strength and art.

(Overleaf) Orca at home. Photo credit: Orca Network.

Orcas—complex minds, complex cultures. Photo credit: Fred Buyle.

Blackfish, as orcas are also known, are midsized whales. Males are usually around eight thousand pounds, but can weigh up to twelve thousand pounds and extend as long as twenty-six feet from rostrum (snout) to tail. As orcas approach adulthood, their iconic black dorsal fin begins to grow and straighten until it reaches its final breathtaking height. A six-foot-high black fin slicing through blue seas is an awesome sight; a group of these towering sails even more so.

In many ways, the rhythm of orca life is much like our own. Courting, coupling, and birthing occur throughout the year. Young females begin their journey into womanhood as teens. Although smaller, females outlive males by decades. On average, female orcas live to fifty, but they are known to live more than one hundred years. One reigning grand dame of the Pacific Northwest's Southern Resident orcas, "J2," is estimated to be more than a hundred years old.[2] She was already a juvenile swimming in crystal bright waters of the north when the world was about to fall into the grip of modern history's bloodiest war, World War I. Males reach sexual maturity a bit later—in their early twenties—and reach twenty-nine on average with the potential of living up to sixty years of age.

But there is so much more to orcas than a litany of biological facts. Every avid orca watcher, enthusiast, and biologist can attest that the still waters of an orca mind run very deep. Hence when we try to hazard an orca's answer to "Who are you?" the task is far from simple. But with neuroscience's understanding of a species-common model of brain, mind, and behavior, and a reminder from Bowlby and the bears that our senses of self are cultivated in relationship with others, we can approach the notion of orca selfhood by looking into how and with whom they live. In the language and concept of attachment, a self is not describable in the denomination of one but in terms of those with whom we develop.

The sense of self emerges from the ebb and flow of interpersonal connections experienced from inception through adulthood. This is implicit in your recognition of "I" or "me." Directly or indirectly, every one of your answers to the question "Who are you?" relates you to someone else, be they an individual ("Barry") or a group (plumbers), or a place (two doors down). We know who we are by those with whom we are associated and also by those whom we are not. "I am female (not male)" or "I am Swiss" (not American). I am a plumber (not a farmer), and so on. Because the relationships that inform a developing self are themselves embedded in a broader, social milieu, a sense of self implicitly comprises cultural values. Thus an orca's sense of self, when developed relationally within the shell of normal orca society, embodies the rules of orca society, which include an essential set of morals and ethics, *les regles du jeu.* A Southern Resident orca's sense of self and values, for example, are part of the fractal that is Southern Resident culture and are distinguished from those living within the neighboring Transient society.

A relational understanding of the self is somewhat new in the modern West where, historically, an individual has been envisioned much like a pinball that goes through life knocking about and interacting with other pinballs, with the space in between being exactly that—empty air through which we move and exchange. A group was understood to be made up of multiple individual human selves that added up to a collection of many. But members of traditional indigenous cultures experience selfdom and the space between self and others in very different ways. Although there is significant variation, for many tribal cultures, values and one's sense of self are collective and interdependent, as opposed to individualistic and independent. This contrast is illustrated beautifully with examples from the Quechua of the central Andes where the self does not stop at the corporeal boundary, nor does it exclude the nonhuman, natural world.

Justo Oxa, a Quechuan schoolteacher in Peru, describes his traditional culture in this manner: "The community, the *ayllu,* is not only a territory where a group of people live; it is more than that. It is a dynamic space where the whole community of beings that exist in the world lives; this includes humans, plants, animals, the mountains, the rivers, the rain, etc. All are related like a family. It is important to remember that this place [the community] is not where we are from, it is who we are. For example, I am not from Huantura, I *am* Huantura."[3]

Marisol de la Cadena, a professor of anthropology at the University of California, Davis, who has lived with and talked with the *runakuna,* the Quechuan people, for over a decade, expands on Oxa's description.[4] "In *ayllu,* humans and other-than-human beings do not only exist individually for they are inherently connected, composing the *ayllu* of which they are part and which is part of them—as a single thread in a woven cloth is integral to the weaving, and the weaving is integral to the thread . . . Humans, plants, animals, and earth-beings, are not from a place[,] rather their entangled relationship *takes place in the sense that as it happens it occupies a space.* Rather than being instilled in the individual subject, the substance of the humans and other-than-humans that make an *ayllu* is their place-making temporal-spatial co-emergence with others."[5]

Three hundred miles away as the crow flies, Canadian anthropologist Inge Bolin found similar cultural patterns among the Chillihuani Quechua who live in rarified Andean heights at 16,500 feet above sea level. Children born into this seamless tissue of life exist simultaneously in the singular and collective, much in the same way that quantum physics describes particle-wave duality.[6] For the Chillihuani, "the earth is alive with its rocks, springs, lakes, meadows, and all other aspects of nature which deserve respect in their own rights."[7]

For most Westerners, the portraits painted by Oxa, de la Cadena, and Bolin may be a bit startling. The idea of growing up in a kind of psychological soup where one's intimate sense of self is shared with someone else, let alone some*thing*—a relatively boundless piece of earth—pushes all kinds of buttons. Modern upbringing prides itself on the cultivation of "I am a rock" individualism, control, and independence wherein fences make the best neighbors.[8] Not so in the Andes. Here people consciously invite an inextricable intertwining of self, sun, and soil.

The Quechua "model" of self provides an attractive way to explore orca perception and experience. When examined in detail, we find that orca development aligns well with that of traditional Andean communities. To begin, we take the lead of dolphin researchers who suggest that such undertakings require "understanding who the dolphins are in their own world and what dolphin-specific meanings are ascribed to various aspects of their surroundings, both natural and artificial, within their own societies."[9] To understand an orca self means understanding orca culture, and this requires a kind of archaeological unearthing to carefully brush away the intricate social layers that make up orca society.

It was not too long ago, a mere decade or so, when mention of cetacean culture raised academic eyebrows. But now whale culture is fairly well accepted across disciplines.[10] Scientists concede that, similar to elephants, orcas exhibit "behavior patterns shared by members of a community that rely on socially learned and transmitted information."[11] The source of this "learned and transmitted information" is found far back in time. For thousands of years, *Orcinus orca* has carved its arching signature into ice-green waters that are whipped by whistling winds and edged by ice-floe-patrolling polar bears. They are a species comprised of dozens of ancient marine civilizations with matriarchal lineages as long as the ocean within which they live is deep. Each individual orca is embedded in an intricate social collective. Biologists have tried to untangle these complex social patterns by breaking them down into discrete organizational packets.

At the global scale, orca populations are grouped into ten ecotypes that are split between the hemispheres. In the southern ocean there are Type A, Pack Ice and Gerlache Type B, Ross Ice Type C, and Subantarctic Type D. North of the equator, killer whale populations divide between oceans—the eastern North Atlantic Type 1 and Type 2 and to the west in the Pacific, the Resident, Transient (also known as Bigg's), and Offshore. For the most part, orca ecotypes relate to specific regions of cold waters, but on occasion, the species does venture into tropical waters. One Antarctic orca tagged "Type B" trekked five thousand miles north to the tepid waters off Brazil where he stayed for a month and a half.[12] Researchers were puzzled at what seemed an extensive, pointless detour. For instance, no noticeable break in pace was observed that could be associated with family matters or special feeding. But upon his return, the blackfish had shed the thick yellowish diatomaceous or algal coating that often accumulates in Antarctic waters. Perhaps, the scientists hypothesized, the quick trip to a tropical clime was a kind of "periodic maintenance migration . . . with warmer waters allowing skin regeneration without the high cost of heat loss: a physiological constraint that may also affect other whales"—a kind of orca spa treatment.[13]

Thanks to joint efforts by the Orca Network located on Whidbey Island, and the Center for Whale Research, Friday Harbor, Washington, the Pacific Northwest orcas have provided scientists with tremendous insights into orca social organization and natural history. The Orca Network, co-founded by Howard Garrett and Susan Berta, is a nonprofit "dedicated to raising awareness of the whales of the Pacific Northwest, and the importance of providing them healthy and safe habitats."[14] Howard and Susan's more than twenty years of meticulous observations, combined with a community-based sighting-information network, have helped make the Southern Resident orcas the best-studied group of orcas in the world. Their brother institution, the Center for Whale Research, is exactly that. Ken Balcomb, executive director and co-author of *Killer Whales: The Natural History and Genealogy of Orcinus orca in British Columbia and Washington,* who, with his organization, has been dedicated to "the study and conservation of the Southern Resident Killer Whale (Orca) population in the Pacific Northwest," is Howard's sibling.[15] A detailed compendium of demographic and family patterns has been created using data collected from field studies starting in 1976 that identified each individual every year. Together, through their care for orcas, and their skill in understanding them as sentient neighbors, not merely data points, these organizations have given us a way to know each and every member of the Southern Resident killer whale community as an individual personality, and have offered up not yet another dry, spare academic paper, but instead a rich, informed archive of orca social organization and diversity.

Neither ecotypes nor their subgroupings mix with each other. Although they may overlap spatially, orcas retain respective nation-state boundaries diligently. In fact, in some cases, ecotypes are sufficiently distinct to be considered separate species or at least sub-species. The Salish orcas, for example, are made up of three North Pacific ecotypes—Resident, Transient, and Offshore. Residents roam the coastal waters off the Pacific Northwest. They earned their moniker because, for the most part, they stay around the waters of Washington State and southern British Columbia. In contrast, Transients, though they share similar waters, typically roam as far south as California and as far north as Alaska and hang out in just a few numbers of two to six. Their family constellations are less consistent, they vocalize less (especially when hunting), and they eat only marine mammals. Residents dine on fish, and fish alone. The two ecotypes do not fraternize but, much like opposing lanes of traffic, go about their own demanding business of finding food. Less is known about Offshore orcas. It seems that they usually travel in relatively large groups of twenty-five or so in continental shelf waters. They may journey as far out as hundreds or even thousands of miles and seem to be mainly fish eaters. Analyses of mtDNA data suggest that Offshore populations may include founding matrilines (lines of common descent from a particular female traced through individual orca mothers, across generations) for Resident orcas.[16]

Their geographic fidelity, the regular occupation of a fairly circumscribed region, suggests that, similar to the indigenous Quechua, Resident orca identities are place-based. The same might be said for Transients, but, like nomadic Bedouins of the Middle East, their identities and cultures may be more expansive, reflective of their extensive and fluid range. As Howard Garrett and Susan Berta remind us, it is essential to not overlook the strong influence of culture on orca life. Killer whale fiefdoms are determined not solely by geography, but also by culture. "Each ecotype lumps several culturally distinct communities on the basis of similar diets and each distinct community or population within each ecotype is genetically unique," Garrett explains. "'Ecotype' implies that orca behavior is set by their environment, but when you look closely, it's really set by their interpretation of their world, including environment, other communities, availability of their chosen prey, preferred group sizes and ranging patterns, etc."[17]

Such culturally imposed, geographically unenforced separation has caught the attention of evolutionary biologists. Orcas are appealing candidates for what is referred to as sympatric speciation, that is, genetic divergence in the absence of any physical barrier such as an ocean or mountain.[18] Generally, two groups of organisms are considered separate species if they do not interbreed and cannot issue fertile progeny. The idea is that speciation occurs when new biological species become separated and form independently through the process of evolution.

The concept of long-term intentional sympatric separation has irritated scientists for scores of years. Some say that yes it happens; others claim that the evidence is not conclusive or that data lag theory.[19] Nor has there been, for many, what is considered a sound biological reason for such "unnatural" division. Phillip Morin, a Southwest Fisheries Science Center cetacean biologist, asserts that orcas have become a poster child for the process "because there are multiple genetically distinct populations [which have not yet been formally described as separate species] with different prey preferences in the North Pacific and Antarctic."[20]

To plumb this riddle, researchers conducted a study on North Atlantic orcas. By comparing the traces of modern and ancient orcas, the scientists thought they might resolve whether two populations off Greenland and Norway would show sympatric speciation.[21] Perhaps different food choices—fish versus marine mammal—across many generations might eventually lead to speciation. Fish eaters breeding with other fish eaters would produce fish-eating progeny, and seal eaters mating with seal eaters would beget seal eaters, and so on. By comparing isotopic ratios and DNA extracted from killer whale bones and skin tissue, the researchers assembled the elements of orca genealogy.

Much to the disappointment of sympatric speciation fans, the analyses did not show that genetic divergence and an associated creation of two separate species had occurred. This may indicate that low genetic diversity derives less from taxonomic separation than matrilineal fidelity—culture may be a "strong evolutionary force" for these whales.[22] Furthermore, the researchers discovered that the East Greenland orcas who were thought to be unyielding fish eaters also ate some seals. The stomach contents of the orca family of six who had been killed by Greenlander hunters had eaten seals even though their teeth showed the considerable wear of herring eaters. (The constant flow of tough herring scales against orca ivory wears the teeth down.) But similar to Offshore orcas who munch on sharks, worn nubs are sufficiently effective for catching and eating their prey. (Seal and sea lion orcas need every bit of their sharp teeth to grab, puncture, and tear tough seal fur and skin that does not abrade teeth.) In contrast to North Pacific populations, then, North Atlantic killer whales appear much less fixed in their food preferences.

Orca supporters were also disappointed with the scientific news. The International Whaling Commission does not monitor orca killing, nor are orcas listed as endangered or threatened, with the exception of the Southern Residents, who were declared endangered in 2005. The identification of more orca species, however, could increase chances for protective listing. Orca status remains "data deficient," because speciation is undetermined. Still, according to the North Atlantic researchers, these are early days, and time may tell. Self-imposed prey-preferential separation may be the first signs of speciation.

There is even more social complexity within ecotypes. For example, within the Resident ecotype, there are three populations: the Alaskan (with approximately 500 individuals), the Northern Residents who patrol the Vancouver Island archipelago (250 individuals), and the Southern Residents who occupy inland waters of Washington State and roam up to Southeast Alaska

Orca pod, Puget Sound, Washington. Photo credit: Orca Network.

and down to Monterey, California. Fewer than 84 individual Southern Residents remain. There's more.

Three pods make up the Southern Residents—pods J, K, and L. A pod consists of a group of matrilines. Pods tend to have no more than at least three matrilines, but the Southern Resident L pod has twelve. Matrilines within a pod are related by a common maternal ancestor and often span five generations. Pods J, K, and L make up the J clan, which, in turn, belongs to the Southern Resident community, one of four communities (southern, northern, western Alaska, and southern Alaska) that in turn belong to one of the ten orca ecotypes, North Pacific Resident. When all three pods unite, it is referred to as a *superpod.*[23]

So, when killer whales are born, their developmental contexts are nested, like *matryoshka* dolls, in layers of complexity. Orca selves mature in a web of interleaved social, ecological, and cultural patterns, wherein the species' entire history, DNA, and experience of plurality live on in each and every individual. In this system, mind, self, culture, and ecology are co-evolved processes in dialogue through epigenetic interaction. As the sharks and chickens teach, each person is yet one more link in an endless chain through time and space.[24] Similar to the Quechua, orca "I-ness" is "inherently relational, a fractal person [who] is always already with others."[25] From birth onward, an orca is "never a unit standing in relation to an aggregate, or an aggregate standing in relation to a unit, but always an entity with relationships integrally implied."[26]

Male orcas typically find a mate outside their pod, and, five hundred days after conception, a beautiful calf is born tail first. Calves stay close to the mother, nursing, for a year, but remain with her for the rest of her life embedded in the innermost circle of the matriline composed of a matriarch, her daughters and sons, and her daughters' progeny. As many as seventeen members have been counted in one matriline. Teeth start to appear around three months of age. When deemed ready, the young orca begins to swim and cavort with other water-breaching, body-hurdling, rowdy youngsters.

Natal relatives maintain very strong bonds and they are never apart for more than a few hours. With two exceptions—L98 (Luna) and A73 (Springer)—there is no record to date of permanent departure from Resident matrilines. Members simply do not disperse. They never leave their social home and associate exclusively with other matrilines within their pod and community. Matriarchs are the centrifugal unifying force around which orca society revolves. As Howard Garrett explains, "When a great grandmother dies, the sister remains in the same pod. Sisters still continue to associate and their offspring may start new matrilines within the same pod."[27]

Such social conservation raises the question of potential inbreeding. Biologists regard orcas as a fission-fusion society and there are rules about who can mate with whom. Mating tends to occur across pods and when there are insufficient mating males, the mating has occurred with associated orcas but not within the immediate family. Genetic analyses using microsatellite DNA indicate that "resident males nearly always mate with females outside of their own pods, thereby reducing the risks of inbreeding." Researchers have also posited that the orca gene pool is so robust and redundant that even if there are double recessive genes, they are overridden by healthy pairs, thereby avoiding potential disabilities.[28]

This cultural constancy can create vulnerability. Compared to other species such as the tiger shark, who variously munches on teleost fish, sea snakes, and turtles, many orca diets are extremely restricted.[29] Take, for example, the Southern Residents. The oil-rich Chinook salmon (*Oncorhynchus tshawytscha*), prized by fishing enthusiasts and human diners, makes up 80 percent of the orca's diet—a cross-species commonality that has caused a major food war.[30]

Before the onset of intensive fishing and dam construction, Chinook salmon ringed the Pacific coast from southern California north to Alaska and around to Russia and Japan. They are anadromous: born in freshwater, they spend most of their remaining lives in the ocean. There are two Chinook "races," stream-type and ocean-type. Compared to their ocean counterparts, stream-types stay in the freshwater in which they are born longer before turning tail and heading downstream to the wide ocean. Ocean-types head back to salty waters much sooner, within the first three months after hatching. They migrate as well, but stay closer to land in estuaries and coastal areas. Chinook swim afar thousands of miles before coming back to their natal streams in spring or summer. The fish is the largest in the salmon genus, with some growing as big as humans, and can weigh up to eighty pounds, a size that garnered them the nickname "king" salmon and "June hogs."[31] Grizzlies and eagles also fancy Chinook.[32] Now, however, prey and predator are equally endangered.

Water diversion and overuse, intense fishing pressures, a multitude of dams that block access to natal streams, and overall habitat degradation have gutted Chinook populations. The Chinook crisis is alarming for orcas in the short and long term. From 1995 to 2001, coincident with regionwide drops in Chinook numbers, the Southern Resident orca numbers fell by 20 percent to an unsettling low of just seventy-eight individuals. Recovered bodies showed signs of starvation. Even though other fish were available, the orcas did not deviate from their traditional foodstuff. This means that the "human-caused drop in Chinook numbers has put the Southern Residents at extreme risk of extinction" and even though orcas "are starving, losing their family and babies, they are not shifting to other fish, like greater amounts of chum or some other species. It is both a mystery and witness to the power of orca culture."[33]

The situation became so dire that the Chinook were listed as endangered species and the Southern Residents were declared an endangered Distinct Population Segment of a sub-species of *Orcinus orca.*[34] But this did not settle well with everyone. In Spring 2015, the National Oceanic and Atmospheric Administration (NOAA) initiated a review in response to a petition to delist Snake River Fall Chinook as threatened under the Endangered Species Act, but the following year, federal authorities denied the delisting. Little progress on salmon populations and habitat restoration has been made. The ship and boat traffic, contamination, noise, and heavy fishing pressures that keep the orcas at precariously low numbers have yet to be reduced.[35] So "if we really want to have healthy ecosystems with salmon and whales in the Pacific Northwest future," Ken Balcomb asserts, "we must all write to our elected representatives in support of a Presidential mandate that returns the Snake River ecosystem to its natural state. When the orcas and salmon are gone, they will be gone forever."[36]

Scientists speculate that there may be some key nutrient in Chinook salmon not found in other fish or marine mammals and, warns Balcomb, "resource availability may be an important determinant of social network structure."[37] As a result, "given the central importance of the social network for population processes such as the maintenance of cooperation and the transmission of information and disease, a change in social network structure caused by a change in food availability may have significant ecological and evolutionary consequences."[38]

The sheer strength of orcas' interpersonal relationships has kept their social structures stable under the pressure of such pervasive stressors. Killer whales have been able to hang on as long as they have, because, Garrett maintains, they adhere to custom: "Orcas are creatures of rules. They act according to very deeply engrained traditions. From infants to adults, they exist in the envelope of their cultures all the time and are tuned in to each other all the time. They are always in acoustic contact, talking back and forth. We see the matriarchs giving commands and guiding family. They are so well connected that in an instant, they can all change behavior on a dime . . . One minute they are headed out, then the next, they have turned direction, synchronized. There is more than body language and behavior involved. Somehow they are making and sharing meanings with each other and the group that convey very sophisticated sensory and cognitive information and sense. Their sensitivity and keen intelligence are astounding."[39]

Orcas such as the Alaskan and Northern Residents who eat herring and cod, in addition to salmon, show greater flexibility. Others, like those of the coast of South America, also are malleable in their tastes in a different way. This pliancy may not only buy them some time before population collapse, but also give them some payback from the humans who have stolen their food base.[40]

The story begins with toothfish, more commonly known by their market name, Chilean sea bass. Toothfish are dwellers of the deep, large-bodied (up to 150 kilograms), long-lived (up to fifty years) and, as members of the cod icefish family Nototheniidae, live between three hundred and 2,500 meters below the surface, in the frigid circumpolar waters. To stay in such deep, cold waters, they have evolved a wonderful system of survival. Their blood contains a glycoprotein that prevents ice-crystal formation in waters at or below the freezing point, keeping the blood liquid. This allows the "white crocodile fish" to swim quite comfortably in waters where few dare to go. They are also the only known vertebrate who does not require hemoglobin, because southern polar waters are highly oxygenated: toothfish absorb oxygen directly through their skin.[41]

In the 1980s and 1990s, a fishing boom brought on a toothfish crisis.[42] Legal and Illegal, Unregulated, Unreported (IUU) fishing depleted populations. Catching the Patagonian and Antarctic toothfish is so lucrative that it is competitive with the elephant ivory trade, another form of pirated treasure.

The Sea Shepard Society, the smart marine watchdog group, reports that "a 1,500-ton catch, not uncommon for one vessel, can be worth as much as $U.S. 83 million."[43] When in 1996 a major toothfish industry was established in Patagonia, local killer whales lost a dietary staple. Soon, however, the resourceful orca learned how to nip off the toothfish who were caught on fishing longlines and take back what the fishermen stole. Needless to say, humans are never good at taking a joke and have fought back with lethal means. Even so, sperm whales have begun taking a page from the orcas' book and are starting to glean toothfish from longlines.[44]

Whether working together to solve a problem or just living communally, orca social life is insular and inward looking, especially among Residents. The young remain protected and cared for within this prosocial cocoon. As Garrett explains, orcas operate by persuasion without physicality: "They are above emotional outbursts and unconscious destructive acts. Orcas don't fight, they don't joust, no squabbles or demands for authority, and they don't kill each other. They care for each other, and teach their children this understanding and respect generation after generation. Orca cultures are very, very old cultures."[45]

Garrett's description and those of other orca biologists are very similar to that offered by neuropsychologist Darcia Narvaez, in her discussion of humanity's ancestral foraging small-band-gatherer-hunter societies, which were characterized by "egalitarian, voluntary sharing of activity, resources, and company."[46] The observations by orca researchers also have remarkable parallels among the Quechua as described by Inge Bolin:

> Chillihuani children are loved and appreciated within the immediate and extended family and the community. They are never left alone, but are always with a caregiver, the . . . children and youth . . . , despite great material poverty, are polite, responsible, compassionate, curious, adventurous, and courageous even at a very young age. I have never seen a child being beaten or treated roughly . . . Children participate in the adult world . . . The competitive element does not rule . . . Children participate in the

> struggle for survival where they make important decisions and master tasks by themselves . . . These young children must have a good understanding of the behaviour of their animals, the terrain they are passing through, the regions where their family has rights to pasture, the climate and seasons, animal diseases and birthing, predators, and more.[47]

In the socio-marine developmental niche in which a young orca is reared, pod members "set the pattern for emotion systems and the nature of their strength and connectivity. When properly developed, emotions facilitate adaptations to the environment, generating a sense of wellbeing, a sense of competence in the face of threats to survival, and a secure attachment, all of which contribute to thriving and the development of a coherent self."[48] This *engagement ethic* develops in the case of tribal indigenous and orca social systems and when "predominant, there will be strong, close relationships with others, and feelings of empathy will be more accessible than feelings of anger or contempt. A supportive early social environment enables [the] individual to tolerate and understand a range of feelings in themselves and others without triggering personal distress."[49] Getting back to the brain-speak introduced in our discussion of grizzly attachment, orca behavioral displays of affection, bonding, care, and attention correspond to "the fronto-limbic autonomic cortico-subcortical circuits of the emotion transacting attachment system," which are areas of the brain known in mammals for orchestrating an engagement ethic. [50]

An orca is never left alone. Males remain by their mother's side for her entire life, and stay with their siblings after she dies. Similarly, Chillihuani elder Don Robert Yupanqui Ccoa insists, "We, the people of Chillihuani, BELONG to our village, just as we BELONG to our families. No one may ever be left out, because loneliness is painful."[51] Only a few exceptions to these orca rules have been observed. Garrett recalls the young male L87 who was orphaned and did the unheard of—switched pods. "In early 2005, L87, known as Onyx, lost his mother. Orphaned males with no siblings sometimes gravitate across pods and attach to an older female. L87 first went to a different K pod and stayed with K7, almost 100 years old. However, she died three years later, so L87 went to next oldest female, K11, about 70 years old. In 2010 *she* died, so he gravitated to 100-year-old J2 and has been there for four years. This cross-pod pollination was previously unheard of. No orcas had ever been [observed] mixing [within] switched pods. It was clear that his change of companion was . . . intended to find the most available grandmother. One theory is that the older females function like a kind of dating-mating service."[52]

The experience of L87 demonstrates something else. Even though orcas abide by strict rules of segregation, they still welcome outsiders from other pods to the community. Pod hopping isn't all that easy and L87 probably had to learn one or two new dialects. Recordings of infant and young orcas indicate that they begin with a basic vocabulary, then learn more as time goes by and as they take on more responsibility for everyday orca chores within the family.

Orcas produce sounds by forcing air through nasal passages. The stream of sound is shaped and directed by the distinctive melon-shaped bump on orcas' foreheads. There are three known communication modalities with both low and high frequencies: echolocation clicks, tonal whistles, and pulsed calls that are used for the various tasks of life. Calls are frequently heard and are used when socializing, finding food, traveling together, navigating, and as orca aural touching when out of visual range. Orca echolocation is also complex. A slower stream of clicks (lower frequency, greater wavelength) is designed to work well for long-distance communications. Rapid clicks (higher frequency, lower wavelength) are appropriate for more complicated messages sent over short distances (just a few meters).

As with other aspects of orca life, communication is highly structured and tied to social organization. Orca clans are acoustic groups of pods with a shared dialect, much like their traditional human neighbors, the Tsimshian.[53] Language is the mortar that binds orca social edifices and establishes the contours of the complex cultural architecture of clans, pods, and matrilines. Each community of Pacific Transients and Residents vocalizes using different repertoires of acoustic signals. Within a pod, matrilines have similar dialects, that is, vocal signatures reflecting relatedness within and across social groups. They also use different modalities of language, depending on the need at hand. For long-range foraging and extended travel, Southern Residents use a whistle, whereas Northern Residents whistle predominantly in close-range social exchanges. Pulse calls are not shared between clans.

Killer whale culture is remarkably flat, which means it lacks any pronounced hierarchical, power-informed ordering. There is only one recognized individual who is believed to supersede others, but only in terms of knowledge: the matriarch. Her authority is linked to age and reproduction, which brings us to another interesting feature of orca cultural biology. Orca females are menopausal and remain fertile until their forties. So far, scientists have identified just three species in the animal world with these hormonal rhythms: orcas, pilot whales, and humans. Although some individuals may cease to bear young because of frailty, disease, or other physiological compromise, females of all other species generally remain fertile throughout their lifespans. Menopause presents a logic conundrum in the eyes of evolutionary biology.

According to the rules of genetic engagement, menopause poses a liability, because it takes a female's future investment out of the gene pool, something that evolutionary biologists consider a substantive disadvantage. If your genes are no longer in the running, then you and your lineage will tend to lose out. In terms of numbers, orcas seemingly don't have a lot of chances to ensure genetic continuity. Females bear only three to five calves over their entire life, and calf mortality is high. Almost half of all young die within the first year. Young male deaths are particularly high. But as usual, nature has alternative ways of making sure that life is democratically proportioned.

A female's shift from reproducer to helper may confer advantages to her genealogy in other ways. Similar to women in traditional matriarchal human societies, middle to elder orca matriarchs assume leadership when they are no longer able to bear calves. Drawing on decades of experience and learning passed from generation to generation, the matriarchs are keepers of ancient orca wisdom. A study on salmon-eating orcas showed that once females enter the postreproductive stage of life, they step up as leaders of the family. The matriarch is the one who brings the family to rich salmon-feeding grounds. Experience collected over generations, what ethologists refer to as "socio-ecological knowledge," is stored in the minds and bodies of elders.[54] The matriarch, as a moving library of historical information, is vital. She advises where to forage in lean times and who is who in the maze of relationships that comprise a society.

Finding food is not always straightforward and on average, among fish-eaters, each orca consumes between one hundred and three hundred pounds of fish a day. So in salmon-poor years, the matriarch's role becomes particularly pronounced. Her deep bench of knowledge guides the pod to safety and to less-well-known food stores in times of danger and scarcity. Biologists hypothesize that this conservative strategy may offset the loss of input to the gene pool. Mortality rates and reproductive success are both tied to salmon abundance. What is lost in terms of quantity (numbers of children) may be gained by an increase in quality (reducing mortality).

There is yet another aspect to orca-eating eccentricities that puzzles scientists. They don't eat humans. Although a great white shark may mistakenly bite a surfer or swimmer, then spit him out, orcas make no such error. With the exception of only one episode in 2014 (and as we shall learn, in cases of captive-held individuals), there is not a single incident recorded of an orca doing anything to harass or harm a human in all of known human history, which includes the thousands upon thousands of years of aboriginal past from waters stretching from the Antipodes, to North America, Asia, and Europe.[55]

For those steeped in neo-Tennysonian tooth-and-claw attitudes toward carnivores, this comes as a huge surprise. Teeth such as those of orcas are extraordinarily well designed to bite, eat, and ask questions later. So why not bother with humans, particularly when humans are so bothersome? It turns out that orcas do bother, but only in certain circumstances, as the story of Tilikum reveals.

Born around 1981, Tilikum is a 12,300 pound, twenty-two-foot-long male killer whale who hails from the Icelandic North Atlantic ecotype. He is the largest killer whale in captivity. When he was first caught, he was sent to Canada to serve as aquarium entertainment. Over the 1970s and 1980s, there was a rash of wild orca captures. *Blackfish,* a documentary that focuses on Tilikum, includes footage and interviews that describe a typical orca cap-

The joy of Orca togetherness. Photo credit: Orca Network.

ture, but the film's hunt and capture scene took place a continent and an ocean west from where Tilikum was caught in Icelandic waters.

Video clips describe the event much like a wild west roundup led by Clint Eastwood's Rowdy Yates. Former diver John Crowe, who participated in the August 1970 orca hunt, recalls the adrenalin rush of pursuit: "It was a really exciting thing to do and so everybody wanted to do it." The film cuts away from Crowe to a high-speed boat cutting through choppy seas. Audio cowboy yelps punctuate the grind of marine motors. It was a no-holds-barred occasion and the hunters left nothing to chance.[56] "Spotters, speed boats, they had bombs that they were lighting with acetylene torches in their boats and they were throwing them in to the water as fast as they could to herd the whales," recounts Howard Garrett.

Just moments after the hunt began, the pod was spotted and the hounds closed in. Boats pushed the orcas into a cul-de-sac where they could be readily netted. The whales swam into a closed arm of the undulating coastline, just as the hunters had planned. But what the hunters did not foresee was that the orcas had their own plan. The pod had silently split into two groups—one made up of the older adults, who stayed the expected course to distract and take the heat while the other group, comprised of young ones and their mothers, skipped off into another channel. All would have ended well for the families but for the predatory aircraft that circled above. Pilots quickly spotted the trick and radioed down the information to the boats, which then changed course and headed out to Penn Cove where they surrounded the natal group. Baby orcas were separated from their mothers, scooped up and whisked away. It was not, however, a success, even by human standards. Five orcas perished, drowned in the process of netting. John Crowe was asked to dispose of them . . . that is, sink the bodies secretly.[57]

As Howard Garrett relates, "The mass round-up in 1970 at Penn Cove on Whidbey Island was terrible. That is where Lolita was caught with six others. There were about a hundred other orcas who swam just outside the nets that were being used to pull the young orcas in. The pod was very agitated. They were calling out and screaming desperately. Everyone could see and hear how distressed they all were."[58] The infant orcas (including the young female who would be named "Lolita" and who has lived in a small tank at the Miami Seaquariam) were spirited away to live in captivity for shows and breeding.[59]

Years after his own capture and that of the other Puget Sound orcas, Tilikum became a household word in 2010 when he killed Dawn Brancheau, a senior trainer at the SeaWorld marine park in Orlando, Florida. Before Brancheau's death, the whale was involved with two other human fatalities: another trainer in 1991, and a trespassing visitor in 1999. In the first death, Tilikum did not act alone. He participated with two female orcas, Nootka and Haida, who were also known to harass and bite the male whale. Trainer Colin Baird believes that the 1991 death of twenty-year-old marine biology student and part-time trainer Keltie Byrne was unintentional: "As best as I can understand it, the three orcas were a little surprised that one of their trainers had seemingly jumped into the pool, although fallen, and they were sort of excited about that, it was something completely out of the norm . . . This wasn't a malicious attack; it was an accident. She had fallen, she had slipped and fell, and taken in a lung full of water. It was not a malicious attack."[60]

But an eyewitness, Nadine Kallen, interprets the death differently, described later by the medical coroner as "due to or as a consequence of forced submersion by orca (killer) whales."[61] Kallen recounts that Byrne "tried to get back out [of the pool] and the other girl tried to pull her up, but the whale grabbed her back foot and pulled her under . . . And then the whales—they bounced her around the pool a whole bunch of times, and she was screaming for help."[62]

Circumstances of the trespassing visitor's death are uncertain (because there are no known witnesses), but there is no ambiguity in the third killing. Dawn Brancheau's demise was videotaped by a throng of tourists watching the "Dine with Shamu" show in which Tilikum and Brancheau co-starred. One YouTube video describes almost two minutes of Tilikum repeatedly grabbing the blonde ponytailed trainer by various parts of her body, pulling her down under the water and thrashing her about. When it appears that Brancheau is getting away, Tilikum again pulls, grabs, swims with, then submerges the woman.[63] The medical coroner's report confirms this sequence. Brancheau succumbed from "multiple traumatic injuries and drowning."[64]

These incidents are not isolated. A similar but nonlethal event occurred in 2006 at SeaWorld when a female orca, Katsaka, also a calf captured off Iceland in the same general period as Tilikum, grabbed wet-suited trainer Ken Peters and repeatedly pulled him down under the water, keeping him submerged for a full minute.[65] Two months before Brancheau's death, another trainer, Alexis Martinez at Loro Parque, was killed by a young male orca named Keto.

Often, orca attacks at Canada's Sealand and SeaWorld in the United States, the two locations where the killer whale caused fatalities occurred, are viewed as accidents in otherwise mutually benign relationships: consider one reporter's view that the brush of the trainer's "ponytail against the nose of a killer whale appears to have been what sparked the whale's attack."[66] The

Brancheau video, however, leaves no doubt about intentionality and is at odds with what has been the pattern of orca-human interactions in the wild for thousands of years.[67] Wild orcas are not known to challenge humans even under the most traumatic circumstances. Howard Garrett described the orca families watching the Penn Cove roundup and capture of young orcas: "There they were, each weighing several tons with powerful jaws and teeth and they did nothing but watch, call, and wait. They were extremely vulnerable. They could have been run over by boats or shot. But they stayed. Witness to their children. They did not save themselves. All they had in mind was the children who were being rounded up. The same thing happened in Japan, twenty-seven years later in 1997. Orca mothers didn't toss the boats there either. The orcas stay within the rules of non-violence toward humans that are so strongly drilled in them. Even in the face of losing their babies."[68]

Similarly, tourists in an Alaska whale-watching group were swimming in the ocean with sea lions thronging the area, when suddenly black fins appeared. One by one, the orcas hunted down and began to feed on sea lions, variously tossing them into the air and chomping down. The sparkling blue waters turned red with seal blood—a rather terrifying turn of events for tourists seeking a peaceful oneness with nature. Notably, amid the chaos not a single human was touched by the orcas. The killer whales accomplished their goal of obtaining a meal of sea lions and disappeared, leaving the vacationers in a bloody, but less than sanguine, state.

Orcas are definitely on task and for the most part, show intense introversion when compared with their sometimes food, seals and dolphins, who are very happy to dialogue and play with humans. Yet as Fred Buyle's experience with orcas illustrates, they are keenly aware of humans:

> We took the Zodiac out just off Senja, Norway's second largest island, way up north at latitude 69N, where two orcas had cornered a school of herring and were gently eating away. We got into the water and they scanned us with sonar and said some codas. After a while they left, but returned fifteen minutes later with a baby! They began teaching the infant orca how to tail slap and catch herring. The baby orca was not very practiced. He got

very close to us and his mother was obviously relaxed about it. A few days later, we swam among an amazing gathering. There were over 100 orcas and a dozen or more humpback whales—all busy herring hunting together. I had seen this before—fin, humpback, and killer whales and big cod eating herring. It appeared that they were not competing but cooperating: orcas taking advantage of the humpback-bubble-encircled herring, fin whales diving in and out. It was as if they had a kind of gentleman's agreement where they all share and enjoy . . . Anyway that day, I ended up about 500 meters from the Zodiac and floating over a depth of 600 meters, alone in the midst of the herring school when I heard an orca. All of a sudden the giant dorsal fin of a massive male appeared. Then there he was right in front of me. I felt him scan me and he started to send me codas. He was looking at me like he was curiously investigating who I was and what I was doing there. I sensed a real restraint and gentleness, as if he was fully aware how intimidating he could be and he did not want to scare me. That was really interesting. I could understand. He seemed to be asking, "So, what are you doing here? Are you here to help me with the herring? Or, are you wanting to eat yourself, or are you just here to watch?" He kept sending me codas, polite and inviting, but I could not answer. He seemed to accept that I was rightfully present, at the herring feast, with all the other whales.

I have to keep emphasizing how much I saw him try to not be intimidating and how he was so mindfully gentle towards me. He was like the reverse of those martial arts masters such as Musashi who, by their internal power alone, can make themselves look huge in comparison with their actual physical size. Yes, that is what he appeared to be doing, but in his case, he was trying to look milder rather than bigger. We spent about ten minutes together, always no more than six to seven meters away. Then a smaller male came, and they left. I felt so honored—and humbled—at his largesse.[69]

Then there is the famous case of "Old Tom." He was a massive adult male orca from the Southern Hemisphere. From the closing years of the nineteenth century through the opening years of the twentieth, he encouraged and led his pod to cooperate with Australian whalers living in the tiny southeast town of Eden. Orcas would swim up the Kiah River to where the men lived and would then breach or make a "flop tail" to convey that humpback whales had appeared. At this point, cued, the whalers would row out, harpoon a whale, and leave his or her tongue as orca tithing. The Davidson family practiced this mutual arrangement for three generations. At times, when an orca got caught up in nets or lines, the fishermen would paddle out to untangle their friend. Everything changed, however, when Harry Silks, a transient man, knifed to death Typee, a male orca, stranded on the beach. Soon afterward, the killer whales disappeared.[70]

Orca restraint and ethics continue to challenge and amaze Garrett:

> Why haven't the Southern Residents changed from their staple of Chinook salmon to some other fish? And why wouldn't mothers attack people who were harming and stealing their young? Pilot whales do some of the same things. There are mass strandings. One goes up on shore and the others follow. A lot of people have proposed that they are disoriented, some chemical contamination that causes this "irrational" behavior. But I think the best explanation is that the pilots are so tightly bonded to their families that it is inconceivable for them to leave the injured one behind. In both the orca and pilot cases there is such a profound social sense of self and culture that [it] transcends what we judge from the outside as some kind of evolutionary *harakiri.* You can see the depth to which these cultural values are engrained in their fidelity to what they eat and how they refrain from violence, except when they must eat, even when their young are in danger.[71]

Having studied sociology and anthropology, Garrett is reminded of human taboo practices. Viennese psychiatrist Sigmund Freud describes *taboo* (the word of which is derived from South Seas languages as *tapu* or *kapu,* rendered into English as "taboo" by Captain James Cook) as "something unapproachable," something forbidden. But it has been translated by others as the "vital principle" that, when violated, undermines and betrays the essential life force.[72] As a collective credo, taboos delineate a non-negotiable cultural boundary. Taboos are prohibitions based on values that are collectively agreed upon by all members of the society. In human cultures, taboos range from cannibalism, to suicide, incest, consuming of animal flesh, and so on. Breaking a taboo is making profane the sacred and brings inevitable social punishment—ostracism, imprisonment, or even death. Critically, a taboo is some act that, if pursued, constitutes a profound insult to the collective group and thereby represents the undoing of the self and one's collective integrity.

Given orcas' finely tuned discernment of physical and social environments, their deeply engrained cultural value to spare human life, and their unwavering loyalty to honor their traditions, Tilikum's violent acts against humans, and Nootka and Haida's interspecies violence toward conspecifics, are astounding. By doing what they did, kill a human and intentionally harm one of their own, the orcas committed what is considered by killer whale society a grave moral violation, if not an outright crime against orca society (much as we humans have crimes "against humanity"). In this light, while the basis for orca restraint against their own kind and humans remains mysterious, what is clear is that capture and captivity can override millennia of orca culture. Whatever Tilikum, Katsaka, Nootka, Lolita (Tokitae), Haida, and others in captivity have experienced has assaulted the bedrock of their orca selfhood.

Prior to his killing of Brancheau, Tilikum suffered a series of psychic body blows: the trauma of his capture, the sudden and profound loss of family, community, and culture, and the chronic stress of captivity, deprivation, and exploitation. By the age of two, when he was captured, Tilikum had completed the second stage of self-nucleation (the first included the in utero period followed by the intense bonding phase of nursing with his mother). He was entering the next phase of self-development where he would expand the complexity and types of his social and ecological—mental and physical—interactions. During this period, a calf begins to spend more time with other members of the natal family, take semi-independent steps into his marine surroundings, and perhaps even hunt on his own.

But, for Tilikum, capture precluded successive phases of teen male friendships and, given what we now know about orca culture, an array of normal relationships and activities that make up orca life at its various scales of social organization. Capture arrested Tilikum's emotional, psychological, social, linguistic, and physical development. Captivity is spare and surreal. He and other captive-held killer whales spend most of their lives in artificial pools not much bigger than their bodies. Tilikum's present pool is a mere twenty-six feet—four feet more than the length of his body. Most whales live alone, with their only interactions taking place while learning tricks from, being fed by, and performing with a human trainer—or being masturbated by a human.

Similar to elephants in captivity, orcas are subjected to intensive breeding programs. Decreased access to wild populations has compelled captive industries to find an alternative source to replenish their stock, especially given the devastating mortality rates that captive life brings. Tilikum has fathered thirteen orcas and has been "in training" for artificial insemination since 1999. It is not certain if the male orca in the twenty-six-second video taken at SeaWorld is Tilikum, but it is a male orca at SeaWorld nonetheless. Holding a plastic bag, black wet-suited SeaWorld trainer Brian Rokeach informs the camera that when the male orca in the pool beside him "sees this bag, he will roll over and present his penis, and, ideally he will emit that sample." The narrator cuts in to say that it has taken months to train the orca. Rokeach continues, "We trained him pretty well to just sit here with us, but certainly when he starts thrusting his tail flukes"—Rokeach nods his head seriously, his left hand grasping the orca's penis—"it is very difficult to hold on." We see a male orca turned over on his back with a pink penis extended. The trainer and another person are holding and what appears to be massaging the orca's penis. The narrator takes over, saying, "This semen collection is part of crucial research into understanding the reproductive systems of some of the biggest creatures in the sea."[73]

From a conventional Western view, it is perhaps difficult to understand how the environment and these experiences could affect something as intangible and elusive as an orca's sense of "self." But we can get an idea by envisioning the self as an embodied being, an integrated mind and body, for, as cultural anthropologist Thomas Csordas puts it, "The body is an existential condition in which culture and self are grounded."[74] Stripping an orca or human of his or her native physical environment, what captivity achieves, involves a kind of psychological stripping that penetrates down to the essence of existence, the experience of self.

Wounds to Tilikum's sense of self were caused by the loss of the structures and processes that conventionally shape a natural orca self, and by those that replaced natural orca socio-ecological development: the barren confines of captivity and exploitive, abusive relationships. Bereft of its natural, protective evolutionary social and ecological tissue, the captive-held mind-body exists in raw nakedness. Those who survive do so by nursing living wounds that never heal. Neurobiologist Debra Niehoff describes how the moral self ruptures from chronic and traumatic stress: "Stress wears away the nervous system . . . Minor insults are seen as major threats. Benign details take on a new emotional urgency. Empathy takes a back seat to relief from the numbing discomfort of a stress-deadened nervous system. Surrounded on all sides by real and imagined threats, the individual resorts to time honored survival strategies. Fight, flight, or freeze."[75]

While unseen to the uninformed eye, the pressures of past and current trauma are ever present, ever poised to erupt. In analyzing the incident, former SeaWorld animal trainer Samantha Berg describes the signs and crescendo: "In the video from the 'Dine with Shamu' show, Tilikum is performing well in the session to start off, but then Dawn 'loses' his attention and he starts giving her mediocre responses to her requests . . . For some reason, Dawn continued to work a session with an animal that was not responding well and with minimal food (fish) at her disposal."[76]

Howard Garrett, who spent many hours poring over official reports, medical records, and speaking with numerous individuals who were either present or very familiar with the circumstances, also notes the build-up to the attack:

> I think a crucial moment occurred that I believe tipped Tilikum into the peak stress that led him to grab Dawn and kill her. The "Dine with Shamu" session with Tilikum had not gone smoothly. In fact, the main show had to be halted due to aggression between

Orca Lolita (Tokitae), victim of capture and captivity, in aquarium show. Photo credit: Shelby Nicole.

the orcas that preempted them doing the routines. So there was tension in the air already, and when Dawn began the "relationship session" with Tilikum one can imagine his relief that for a short time nothing more was expected of him, and he could simply relax and have a moment with the trainer he knew so well. He was routinely dominated and manipulated by her, but for this short moment they could have some quality time together. Considering his life since capture, enduring extreme stress by extreme physical incarceration and emotional abuse from whales

> and trainers alike, this must have been a glimmer of a happy moment, face to face with Dawn, with her saying sweet things and probably stroking his rostrum, [the] nose-like end of his face. Then, after this brief moment of relative calm, he hears the call from the viewing window to end the relationship session and appear on command at the viewing window for his photo session, to be directed to hold a pose and have cameras flash in his face. I believe he simply could not endure the return to being bossed around with no real reward or comfort. I think that's the moment he thought the orca equivalent of "Screw it, I'm not going to end this moment with Dawn and she's not going anywhere." He slowly pulled her in, as shown in the last second of the tourist video, and once he had violated the rules by pulling her in he knew he would be reprimanded with a long time out that would be even more stressful, and he did what he thought he had to do to put an end to the stresses, or he just went into a rage . . . I doubt he knew what a shit storm would follow his attack on Dawn, but he knew at least she would no longer be part of his miserable life, so at least he could have an effect on his own life for a change.[77]

Responding to minor insults coupled with real threats (the helplessness of captivity and aggression by other whales), Tilikum—who was unable to freeze because he was being forced to entertain, or to flee, since he was locked into a small pool—was cornered into using the only remaining option, albeit one absent from orca traditions: to fight and kill Brancheau.

In the words of Emily Dickinson:

> And then a Plank in Reason, broke,
> And I dropped down, and down -
> And hit a World, at every plunge,
> And Finished knowing – then –[78]

Female orcas who are captured or birthed in captivity undergo similar ordeals. After being bought, Lolita, one of the seven infant orcas captured in

Broken teeth, broken souls. Photo credit: Shelby Nicole.

the Penn Cove hunt, was sent to the Miami Seaquarium. Her original name was Tokitae (also Toki), but in a competition run by the *Miami Herald,* she was renamed the oddly unsettling name Lolita, after the vulnerable young seductress of Nabokov fame. She was housed with a male orca, Hugo, captured in Puget Sound, and probably a relative of hers. During an aquarium public show, Hugo slammed his head into the pool wall, breaking the glass, causing a brain aneurysm, and ripping off his rostrum. Although his rostrum was stitched back on, he died from his injuries.

In 2015, in answer to a petition, NOAA determined that Lolita, as a member of the endangered Southern Resident pod, falls under the protection prescribed by the Endangered Species Act.[79] Efforts to bring Lolita back and release her to her home waters continue, despite disagreement about the advisability of releasing captive orcas into the wild. Some cite the death of Keiko, the first captive-held orca who was reintroduced to the wild—and star of the film *Free Willy*—as evidence that reintroduction is not viable. But others maintain that Keiko's release was successful. "While the industry talks about Keiko as a failure, they fail to mention that 17 orcas died in captivity during the time Keiko was being successfully rehabilitated and released."[80] Howard Garrett states: "I am confident that Toki is capable of readapting to her native waters, and that she remembers her family and how to be a Southern Resident orca, and the reason in a word: sociology." Toki's mind and soul will revive when she is once more connected to the wellspring of her life: the Southern Resident orca society.[81]

Physically and psychologically damaged, the outlook for many captives is grim. Orca expert Ken Balcomb states, "Tilikum is basically psychotic. He has been maintained in a situation where I think he is psychologically unrecoverable in terms of being a wild whale."[82] And it appears that the story of Tilikum will be ending soon with no chance of even short-lived freedom. SeaWorld Orlando announced that the great whale is dying from an incurable disease.[83] In captivity, the teeth of this marine mammalian predator who sits atop the trophic pyramid unchallenged often become infected and break from gnawing on steel bars and pool concrete and it is suspected that this is the source of his illness. SeaWorld's president and chief executive officer has also announced that SeaWorld is "ending our orca breeding program."[84]

But such measures cannot cauterize the searing wounds that the great orca society has been dealt. It will take more than a cessation of breeding to right these wrongs. Until humans relent from dominating the oceans, orcas will have to struggle to uphold the ancient beliefs and values that have defined their sense of selves for millennia.

4

Crocodiles: Emotional Intelligence

It is the dry season. Midday heat has made a still life of the Indonesian savannah. The only water sits in shallow pools. Every movement is conserved. As a butterfly alights on a stem of brown grass, a blue-black water buffalo eases his massive body into the cool mud. The stillness is so intense that it is almost loud. Then the day's breath shifts almost imperceptibly—a slow and rhythmic *hiss, hiss* breaks into the hum of insects. Peering warily with one white-rounded eye, the buffalo looks toward the approaching sound and spies a Komodo dragon pushing toward him with measured deliberation. The bull fumbles quickly to his feet. Suddenly, the lizard's gait changes to a rapid stride that mimics the report of a muffled machine gun.[1] The lizard

lunges, and the buffalo parries to no avail. His rear leg is bloodied and he moves away. The Komodo does not follow his prey. He does not have to. His mission was achieved by breaking the buffalo's skin.

Daylight fades into night. Another dawn breaks, another, and yet another. On the sixth day, morning's timid light reveals the buffalo's still body settled in the mud. Flies buzz around his bowed head and the giant lizards begin to gather. One by one they take their places around the expiring bull. By the time the Komodo dragons commence feeding, the buffalo does not even flinch. The silence is broken only by the sound of rending flesh.

Pretty gruesome, but just what we would expect from a giant lizard. After all, reptile brains are, well, reptilian. We expect a cold-blooded killer, devoid of emotion and capable of instinctual response alone, to act like this. Yet according to neuroscientists and herpetologists, this portraiture does not fit the facts. Killers they may be, and ectotherms as well, but lizard brains are not some reduced dimensionality of our own. Underneath the skull beneath the skin, scientists have found someone whose outlook is much more similar to ours. As neurobiologist Erich Jarvis asserts, reptiles have what we have and vice versa: "A reptile brain is similar to a bird brain and both are similar to mammalian brains which implies that reptiles may have parallel capacities to think, feel, experience consciousness, and related abilities to mentally function."[2] Quite a statement, but its source is quite the scientist.

Jarvis is a professor and lab head at The Rockefeller University, in New York City. For more than twenty years, he has conducted comparative studies of bird, mammal, and reptile neuroanatomy to understand the evolution of song and language and our cross-species relatedness. In 2005, he and a hefty list of eminent co-authors published an extraordinary paper in the prestigious *Nature Neuroscience Review* called "Avian Brains and a New Understanding of Vertebrate Brain Evolution." The abstract opens with this intriguing proclamation, "We believe that names have a powerful influence on the experiments we do and the way in which we think."[3]

Gertrude Stein and Pierre Abelard would agree.[4] Names not only tell us what things are; they tell us what they mean to us. Naming is implicitly re-

The emotionally intelligent Alligator. Photo credit: Dr. Marge Peppercorn.

lational, connecting what we see with what we experience. Nominalization defines reality, even if that reality does not correspond to the reality of the named. Such is the case of birds and reptiles, whose mis-naming started more than a hundred years ago.

Ludwig Edinger was a German scientist who lived in the late 1800s. He was founder of the Goethe University Frankfurt am Main Neurological Institute and is known today as the "father of comparative neuroanatomy"—the cross-species study of the brain and its evolution. Edinger fashioned a model of the brain that rose from the culmination of centuries-old theories dating back to Aristotle, when nature and all its inhabitants and elements were thought to exist as a neatly ordered hierarchy, what the Greeks called *scala naturae.*

Carl Linnaeus, Charles Darwin, Herbert Spencer, Charles Lyell, C. G. Jung, and others all took part in this stratified thinking. Brain structures and consciousness were envisioned much like the geologic layers that constitute the Earth's landforms. By comparing details of brain morphology across and up and down the evolutionary tree, neuroanatomists believed that they could trace the origins of the human brain. Present-day brain structures and functions were conceived by Edinger as having evolved in a progressive, stepwise manner from fish, to an intermediate aquatic-terrestrial amphibian, to reptiles, then to birds and mammals, and finally, to nature's crowning glory, humans. Our lineage separated us from ancestral life forms through successive natural selection and evolutionary pressures.

Similar to the body, mental evolution was envisioned as a gradual progression in one direction from undifferentiated homogeneity to highly complex differentiated heterogeneity. Vertebrate nervous systems were divided into six main components: the spinal cord, hindbrain, midbrain, thalamus, cerebellum, and cerebrum. These components were thought to reflect ever greater levels of mental sophistication. Our human forefathers and foremothers acquired increasingly more sophisticated capacities that superseded the minimalist stimulus-response of lowly reptiles. As our brains rose up the evolutionary stairway, they acquired intelligence and consciousness, leaving the humble bacteria and reptiles behind to wallow in the mud of primitivism. At least, that was the story.

For decades, neuroscience was dominated by this stratified, privileged view in the form of Paul McLean's "triune" model of the brain. In his conceptualization, the human brain was made up of a nested set of subsystems: the reptile (R-complex) system, which overlays the limbic system (paleomammalian), which in turn is topped off by the neocortex (neomammalian).[5] All this seemed very logical and, like evolution itself, conservative. But *scala naturae* and the triune brain were ideas built on sand because the foundational assumption of hierarchical ordering was false. Today's model of the brain describes an integrated view of form and function where all three areas described by McLean work in equitable cooperation. Even the elevated status of the cortex, that portion of the brain that sits above it all and where higher-order functions such as cognition are said to generate, is being challenged by evidence showing "that brainstem mechanisms are integral to the constitution of the conscious state."[6]

Edinger and his colleagues exacerbated the errors of their predecessors when they pointed to cross-species differences in the organization of the cerebrum—what is also referred to as the telencephalon—as evidence for phylogenetic ordering. He correctly noted that observed differences, in particular those between mammals and non-mammals, seemed greatest in the telencephalon. He thereby imposed the same ordered pattern of Aristotle's *scala naturae* to telencephalon substructures by using Greek prefixes to name layers denoting evolutionary chronology: paleo, archi, and neo. The nomenclature, however, was not only mixed up (*paleo* was incorrectly affixed to the oldest structure instead of *archi*), but a misinterpretation of the internal brain cell structure and relationships led scientists to believe that birds and reptiles lacked highly developed pallial structures—the smart parts of the brain that we and other mammals enjoy.

What Jarvis and his colleagues discovered was that even though the brains of birds have nuclear and mammals have laminar pallial-shaped neuroarchitecture (brain cell structures), they come from the same ancestor. Brain capacities are conserved despite having different evolutionary trajectories, or to put it another way: "The avian cerebrum has a large pallial territory that performs functions similar to those of the mammalian cortex. Although the avian pallium is nuclear, and the mammalian cortex is laminar in organiza-

Severe habitat loss forces Alligators to use golf courses. Photo credit: Dr. Marge Peppercorn.

tion, the avian pallium supports cognitive abilities similar to, and for some species more advanced than, those of many mammals."[7] Our mammalian cerebrum, including its highly regarded neocortex, the part of the cerebral cortex that is considered to be the most recently evolved, is homologous with that of birds, crocodiles, and other reptiles. Evolution has not made any one species better than the other, just different. Each species is, by definition, expert at what it does.

The neuroanatomical details described by Erich Jarvis and his colleagues help to unravel the scientifically outmoded image of reptiles as primitive, unthinking beasts. Reptiles may appear like back-to-the-future time travelers, but they are as up-to-date as anyone else, and humanity's violence-ridden ways are no more reptilian than the sensibilities of snakes and lizards are specifically human. When celebrated cosmologist Carl Sagan coached American politicians to mitigate terrorism by embracing the cerebral cortex that gave rise to modern "civilization," in lieu of "less reptilian means," he grievously erred.[8] Better a reptile cortex than our own to achieve pacific outcomes.

The same goes for dinosaurs. Lessons from the research of Jarvis and other scientists not only deconstruct privilege in the present tense, they dispel the notion that intelligence increased over evolutionary time. Brontosaurs and pterodactyls weren't less smart than humans or other contemporary species are today. They were able to thrive and survive in conditions that would have killed off most modern-day species. Judged by size and weight alone, *Tyrannosaurus rex* (measuring forty feet long and weighing almost seven tons) was quite successful. The bipedal, tiny-armed, tail-balancing giant lizard did just fine stalking, scavenging, and stomping his way through the animal kingdom until a Cretaceous-Paleogene comet dashed almost everyone's dreams into obliteration.

Since the seminal work of Jarvis and colleagues, other scientists have begun to celebrate reptiles as a "new model for brain evo-devo" research.[9] The new appreciation for reptilian perspicacity augers both well and badly for our scaled kin. The new view carries with it greater respect, and hopefully ethical and legal change that offers protection from exploitation. But it may mean that more lizards, snakes, and turtles will be recruited to join myriad animals—parrots, cats, horseshoe crabs, rats, and so on—who are employed as experimental surrogates in studies seeking to probe the mysteries of the mind and body, human and otherwise.

In any case, crocodiles, alligators, iguanas, rattlesnakes, and their relatives have stepped from the shadows of taxonomic prejudice ready to enjoy the fruits of parity and take their rightful neurobiological place aside humans, other mammals, and birds. But are neuroscientists' predictions congruent with experience? Do ethological observations mirror the patterns that neuroscience predicts? Is there field evidence that reptiles think and feel? The answer to all three questions is yes.

Over the years, with patient and careful observation, herpetologists have gathered behavioral data that match what Jarvis and his colleagues predict. For example, the more than meticulously documented seventy-three behaviors of the mouth-snapping, blood-slurping Komodo dragon (*Varanus komodoensis*) can be related to neural substrates from which our own species draws to revel in the restless seas of emotions, aesthetics, and more.[10] But all this requires a radical makeover of human perceptions and reptilian re-branding is not an easy feat, particularly in the instance of reptilian predators—crocodiles, alligators, and snakes who can eat or poison people. The visage of a crocodile's gaping, tooth-lined jaws stirs deep, dark, archetypal waters. Captain Hook's nemesis is typically caricatured as a swimming Terminator or *Walking Dead* zombie propelled by a naked brainstem topped with a large, overactive amygdala. Nonetheless, as neuroanatomists remind us, crocodiles possess much more than a few rudimentary brain cells, and a more careful exploration of crocodile habits reveals psychological depths that lie behind that scaly, snouted face.

A little information about crocodile taxonomy will lay the foundation for our journey into the depths of crocodile minds. Crocodilia (or Crocodylia) is a complicated and controversial order, and the "lively debate" that has reigned for decades continues. Within Crocodylia, there are three families of crocodilians, all of whom showed up on the planet approximately 83.5 million years ago: Aligatoridae, which is made up of four genera and eight species (two species of alligators and six caimans); Crocodylidae, comprised of two genera (*Crocodylus* and *Tomistoma*) with fourteen species; and the lesser-known Gavialidae family, comprised of one genus and one species, the gharial (*Gavialis gangeticus*). False gharials, now called "tomistomas" (*Tomistoma schlegelii*), who, in the past, were classified with the Crocodylidae family, have been found to be more closely related to the gharials. Gharials (whose name comes from the Hindi word *ghara,* to describe the pot-shaped nodule on their snout tips) are critically endangered freshwater, fish-eating crocodilians.

To the practiced eye, alligators and crocodiles look quite distinct: alligator visages are rounded and blunt, whereas crocodiles are narrower and pointed. The real giveaway is the teeth. When their respective mouths are shut, the crocodile's sharp snaggletooth, its fourth tooth, remains out and glimmering. But of course one might ask, "who's counting teeth around a crocodile?"

Most people will only encounter a crocodile dozing aside a zoo pool's edge, or spy one across tepid waters basking on some shore. But crocodiles are very active; they can and do travel long distances and faster than one might imagine. In addition to species that migrate from water body to water body, crocodiles such as the American (*Crocodylus acutus*) and saltwater (also called the estuarine or Indo-Pacific; *Crocodylus porosus*) are ocean-going.[11] The American crocodile hails from the mangroves, salty cays, and islets of the neotropics on the Atlantic and Pacific coasts, Central America, southern Florida, northern South America, and Caribbean islands, whereas "salties" are found from India throughout Southeast Asia, the South Pacific Islands, and northern Australia.[12]

Fred Buyle describes the American crocodiles he encountered on dives off the coast of Mexico. These individuals make an impressive swimming pilgrimage to a small atoll, where they spend several weeks. Buyle laughingly recalls his first encounter: "At first it is disorienting to swim with a huge reptile because they are so different than a whale or shark. It is strange to dive with a croc because they can be totally still on the sea floor, then suddenly rise with the ability to move in every direction and even backwards—it's such an extraordinary experience when you're used to other large marine animals. But they are elegant and amazing."[13]

Crocodiles are not the only swimming reptile. Diverse sea snakes, sea turtles, and many other scaled species also live in or take to the water, and occasionally the land-bound, seventy-kilogram, almost-four-meter-long Komodo lizard takes to the ocean. It is not uncommon for the giant lizards to swim half a kilometer underwater between the isles of Komodo and Nusa Mbarapu. They do not seem to dive deeper than four meters, and at every one-hundred-meter interval, they rise to the surface, catch a breath, and then submerge to swim on for another hundred meters until they reach their planned destination. They time their voyages well, usually only migrating to sup on the leavings of fishermen and their goats when they can take advantage of northerly winds.[14]

The two alligator species, the American (*A. mississippiensis*) and the Chinese (*A. sinesis;* also known as the Yangtze alligator), live in freshwater riparian zones, marshes, ponds, and similar habitats. American alligators also like to spend time on the Gulf Coast side of the Florida Everglades. The Chang Juang (Yangtze River) and Tai Hu (Lake Tai) were home to great numbers of Chinese alligators until humans drove their populations nearly to extinction.

Recent DNA studies indicate that there may be many more species of crocodiles than has been assumed. For example, by employing "rigorous molecular and morphological species delimitation methods," researchers have found that there are at least seven distinct African crocodile species.[15] As Vladimir Dinets, of the Department of Psychology, University of Tennessee, points out, the Nile crocodile (*Crocodylus niloticus*) is actually split into two species, its neighbors, the dwarf crocodile (African dwarf [*Osteolaemus tetraspis*]) split into three lines, and the critically endangered slender snouted crocodile (*Mecistops cataphractus*) into yet another two species. Future taxonomy may become even more complicated.

Microsatellite data suggest hybridization between the Cuban crocodile (*Crocodylus rhombifer*) and American crocodile (*Crocodylus acutus*).[16] This discovery has conservationists worried because it means that two "Evolutionarily Significant Units" require protection to preserve genetic diversity. As a consequence of overlapping breeding seasons and movement inland to freshwater swampy areas away from human disturbances, male American crocs may be managing to mate with females of the now highly endangered, endemic Cuban crocodiles. This is also happening in Mexico's Yucatan Peninsula with the endangered Morelet's crocodile (*Crocodylus moreletii,* also called the "Mexican crocodile," named after Pierre Marie Arthur Morelet, the first European to identify the species). The resulting hybrid progeny will likely make it more difficult for these species to remain distinct, which may eventually lead to a reclassification of the American crocodile in the Greater Antilles.

For most crocodilian species, a home must offer certain features: both shallow and deep waters, land on which to bask and nest, and the requisite balance of different food types. They are top-of-the-food-chain carnivores in their neighborhood, and, similar to white sharks, their diets vary as a function of species body size, mouth size, age, habitat, and food availability. Depending on these variables, crocodilian food includes invertebrates, fish, birds, other reptiles, and mammals. Nile, saltwater, and mugger crocodiles can reach an impressive four to five meters and carry considerable girth. (The term "mugger" comes from the Sanskrit *makara,* meaning "crocodile" or "mythical monster.")

Hunting styles vary by prey. Similar to many other carnivores, Nile and saltie crocodiles typically capture their meals by ambush hunting. Nile crocodiles wait, submerged, their periscope eyes watching carefully, and when the moment is right, launch a powerful tail-driven assault on a kudu or at times, a lion drinking at the water's edge. Unless it is a long-legged stork who can be handily snatched in one gulp, crocs often pull larger prey to the river floor, where they hold the animal down until he or she drowns. On land, these massive reptiles lie motionless, waiting for the most propitious moment to lunge forward on their short but fast-moving legs to grab their meal-to-be. It therefore came as a surprise when biologists discovered that these world-famous flesh-eaters sometimes dabble outside carnivory, giving themselves a more eclectic diet.[17]

Researchers identified thirteen (out of eighteen species whose dietary information was available, or 72 percent) that partake of fruit, legumes, and grains. This not only changes how crocodiles are viewed but also expands how their role in the ecosystem is understood. With a more diverse vegetarian-supplemented diet, it has been proposed that they function as kind of an oversized reptilian bee. That is, given "the biomass of crocodiles . . . it is likely that crocodilians function as significant seed dispersal agents in many freshwater ecosystems."[18] (Others counter that any such dispersal would be limited or ineffective because the consumed seeds can be rendered inviable by stomach acid.)[19]

In addition to scavenging, taking the leavings of others, biologists have discovered that one crocodile, Morelet's, engages in *kleptoparasitism,* stealing other animals' food. Relative to others, this medium-sized (about three meters long) crocodile, with its distinct silvery brown irises, is considered shy.[20] Although there have been occasions when encounters with Morelet's crocodiles have left humans the loser, the species' diet is mostly composed of smaller creatures inhabiting the land-water interfaces of Mexico, Guatemala, and Belize. It was in such habitat that crocodile researcher Steve Snider found out about the crocodile's kleptoparasitic tendencies.

One day, Snider was paddling down the Macal River in sunny, tropical Belize, when he came upon the body of a freshly killed Baird's tapir (*Tapirus bairdii*).[21] The body was large, weighing an estimated 230 kilograms, and was lying at the water's edge. The tapir's cranium had been "crushed and blood was draining into the water, but otherwise the carcass was intact." The way in which the tapir had been killed and surrounding tracks indicated that the hunter had been the exquisite jaguar (*Panthera onca*). Sure enough, the biologist could hear the big cat in the bushes about twelve meters away making "loud exhalations, grunts, and low, agitated growls." The biologist paddled to a concealed position about fifty meters downwind from the carcass and watched. Guess who was coming to dinner?

A large Morelet's crocodile started swimming rapidly toward the tapir from upstream. Just offshore, he (or she) paused midstream for a half hour, waiting and watching. Everything must have seemed in order, because after the cautionary pause, the crocodile swam over to the carcass, where he proceeded to make "numerous bites on the abdomen, legs, head, and haunches, eventually tearing the abdominal wall."[22] He then took the tapir's stomach and carried it midstream, and "released it before returning to feed inside the body cavity." The crocodile went on digging into the intestines, pulled them out and "shook them violently in the water, seemingly in an attempt to express fecal material, before consuming this organ mass." A smart move; other carnivores such as pumas take the same precaution, since any gastrointestinal system carries a suite of bugs hell-bent on causing havoc if ingested.[23] Earl Showerman, a physician with thirty years of experience in the emergency room, confirms that of the things that go bump in the night inside us, "the list is endless but the top include *Helicobacter* (the nasty bug that causes gastritis, esophagitis and ulcers) that lives in the incredibly low pH environment of the stomach in some people . . . Other, virulent strains of *E. coli, Salmonella, Helicobacter pylori, Shigella,* and *Enterobacter,* and *Vibrio cholerae* (cholera) can cause serious diseases."[24] Although susceptibilities to bacteria differ across species, the idea behind shaking and rinsing the intestines is the same: it is a precautionary way of getting rid of unpalatable feces and/or pathogenic worms and bacteria. Despite their seemingly raunchy eating habits, biologists suggest that crocodiles seem aware of the potential danger and unsavory nature of intestines.

Eventually, after several unsuccessful attempts to pull the remnants of the tapir into the water, the crocodile sank back into the river and disappeared. Later, when night set in, the jaguar returned to finish up the haunches of his hard-won, and now greatly diminished, meal.[25]

Let's take a moment to replay the Morelet's crocodile scene, this time with neuropsychology in mind. The researcher's keen observations, when viewed through this lens, offer us a wonderful opportunity to marry ethological understanding of the crocodile with knowledge of the working brain and mind. The Belizean reptile clearly indicated that something more was going on in his head than crazed, unblinking instinct. There was determined purpose in the crocodile's stealthy ways. Shark, crocodile, bear, and other carnivore behaviors are informed by strict and unyielding rules imposed by living off the hoof. Their food-finding ways have evolved to optimize success in the specific environments in which they live. Reptiles are no different in this sense than white sharks or grizzly bears: all of their brains must be finely tuned to effectively match the particular social and ecological challenges that they encounter and to accomplish the necessary work involved in surviving and bringing up their young. In order to safely procure that dead tapir, the Morelet's crocodile in our example had to do what other crocodiles and alligators must do on a regular basis: integrate 100 million years' worth of genetic memory and intuition with affective and cognitive centers in the brain.

From a psychologist's point of view, the crocodile qualified for high marks in emotional intelligence: a requisite for any hunter. First coined in 1964 by Michael Beldoch, the term *emotional intelligence* became popular through the work of psychologist Daniel Goleman.[26] Until then, emotions were considered unnecessary for solid thinking and decision-making. Indeed, they were regarded almost as contaminants, distractions best left for poets and lovers. Cognition—thinking—was considered to be the only mental function worthy of performing complex tasks and analysis. For this reason, most intelligence assessments such the famous IQ (intelligence quotient) test focused on cognitive skills alone. Intelligence was considered synonymous with cognition.

Then, in the decade of the brain, the 1990s, new insights into brain structure and function began to right centuries of Cartesian wrongs. A new field—affective neuroscience, or the study of the neurobiology of emotions and feelings—reinstated emotions as a valid member of the brain's toolbox and acknowledged them as having an influence on almost everything we do. How and what we feel are integral to how and what decisions we make. Most of the time, when things work out well—meaning that we get what we intend to get—instincts, feelings, and thinking have worked hand in glove. Good relationships and good decisions require a properly functioning system of emotional checks and cognitive balances.

Of course there are still times when the heart overrides the mind or vice versa. We can all recall instances, such as falling in love, when the right circumstances crystallize and our feelings—and their somatic expressions—overrule rationality. As Carnegie Mellon psychology and economics professor George Lowenstein makes the point, "Even when people have a realistic understanding of their own self-interest, immediate emotions can cause people to 'lose control' of their own behavior."[27] But the present case of the Morelet's crocodile is an example of well-tempered affective and cognitive coordination.

Goleman lists four key elements of emotional intelligence: self-awareness (the ability to perceive what and why we are feeling), empathy (knowing what someone else is feeling), self-regulation and management (the ability to handle distressing emotions appropriately), and skills that bring all of these qualities together for appropriate thought and action.[28] Self-awareness is essential for getting what you want, which, in the Morelet's case, meant successfully acquiring and consuming the tapir. If a hunter is not aware of the state and condition of her body and mind, then she compromises her ability to judge accurately whether a particular action constitutes a threat or benefit, death or food. For example, the Komodo dragon could easily have been wounded by the buffalo yet his self-awareness, his deft perception of his capacities, as well as his physical acumen and ability to assess the terrain, led to a successful strike.

Goleman's second criterion of emotional intelligence, empathy, goes hand in hand with self-knowledge. Empathy is not always that warm fuzzy feeling for someone else. We may not think of it as such, but the crocodile's ability to read the scene is a kind of empathy. It entails knowing who that person is and what they are feeling and thinking when they are fully aware that you plan to eat them. Apex predators are kings and queens and generally rule the land or water as the case may be. To maintain their thrones, they must "know thine enemy": they must stay aware of the state and capacity of the opponent, be they prey or competitor. This involves knowing how the minds of their prey work, what they are thinking and feeling down to the minutest detail and how these qualities stack up relative to their own. It also includes being aware of the immediate environs—for example, who or what might throw a wrench in the works.

The crocodile fulfilled the third criterion of emotional intelligence with his ability to self-regulate impulses and emotions. It was a full thirty minutes before the croc flicked his massive tail to kick-start the graceful undulation that would power him to shore. Hunger is a powerful force, and being able to refrain from immediate gratification by dashing over and digging into the juicy meal that the tapir promised suggests what neuropsychologists refer to as good affect-regulation abilities. Effective and appropriate emotional regulation implies that the crocodile's higher-order executive functions were in excellent working order and were not overrun by an overly excited limbic system. This capacity is usually attributed to a well-balanced relationship between the cortex and subcortex regions of the brain—something, the bears and orcas taught us, that is cultivated early on via a secure attachment.

Instead of thoughtlessly diving into a mouth-watering, readymade meal upon detection, the crocodile approached the tapir with studied caution. Restraint is essential to hunting success, whereas emotional excess has a cost. As we saw in the case of the water buffalo taken down by the Komodo lizard, any lapse in knowledge or judgment, or any overflow of emotion, can easily lead to injury that, in the wild, can rapidly prove fatal. When the croc was ready to approach the tapir, the combination of self-awareness and emotional regulation led to the perfect physical move. Crocodilians are built less for endurance than for "an intense burst of power designed to overcome the victim's resistance quickly"; in particular, they prefer to employ their famous "death roll" and hold their "feebly struggling prey underwater until it drowns."[29] The crocodile's élan might suggest that it is easy for him to tackle and overcome a muscular herbivore, but victory is no small feat. Every decision brings vulnerability, so committing to any given action is serious business.

These details are easy to overlook from the vantage of our twenty-first-century society. The highly controlled elective lifestyle of modern, urbanized humans affords all kinds of sloppiness and inattention. Unlike a white shark who must grab with precision, we can put off gathering food for a few days because we have a fridge full of ready-to-eat food. Unlike the grizzly who must get to work as soon as possible to dig a den that will withstand the brutal elements of the north, we can delay because there is a device on the wall called a thermostat. Unlike the puma mother's fierce defense of her babies from wolves, most pushback from an emotional outburst among humans stays verbal.

Finally, the crocodile showed that he was able to assemble all the necessary data derived from sources both internal (subjective information communicated via his limbic system) and external (objective information about the lay of the land, jaguar and so on) and act on this input profitably. He was charged with gauging the appropriate strike strategy and with having a measure of confidence, a sense of self-efficacy, to follow through once the prey, albeit dead, was tackled. There was a reason that he sat out in the water midstream for a full thirty minutes. His sustained, careful surveillance bought time for a process of evaluation that might have involved considering, consciously or unconsciously, various possible scenarios—trying out some gedanken experiments, playing through some "what ifs."

Waiting has its advantages, but it has its disadvantages, too. If the Morelet's crocodile had hung back too long, the jaguar might have returned to retrieve his prize or some other hungry crocodile might have disrupted his best-laid plans. If the biologist heard the jaguar, then we can be sure that the crocodiles did as well. If an encounter with another predator were to occur, then even if our crocodile won in the end, he might be wounded, something that predators seek to avoid at all costs, since a delicious meal is worthless if you get badly injured or killed in the process. This detailed portrait, then, shows an emotionally and cognitively finely tuned hunter, able to weigh all of these considerations and come up with the perfect plan.

There are additional, admirable features of our scaled friends. Crocodiles maneuver in the water and on land. To do so requires sophisticated liminal physiology and breathing abilities that are the result of millions of years of evolution. Crocodiles are air breathers, but they respire quite differently from mammals. Often caricatured as giant lizards, crocodiles are more closely related to birds. Unlike us, whose breath ebbs and flows, crocodile and bird airflow is unidirectional, that is, it flows in one direction. When mammals breathe, air moves through nested sets of smaller and smaller airways to the alveoli termini, where oxygen is absorbed into the bloodstream and carbon dioxide is traded back and out. In birds, this exchange occurs along tiny tubes, or parabronchi, that enable them to soar in the air, via a mechanism that is thought to accommodate strenuous periods of flight which may cause hypoxia. Air sacs then might be used as chambers to control flight pitch and roll. A tricky flow path through crocodile lungs creates an aerodynamic valve that accomplishes their own style of air exchange.

Researchers posit that this handy feature might have given archosaurs (ancestors of birds and crocodilans) the evolutionary upper hand after the Permian-Triassic extinction that occurred some 251 million years ago and when 70 percent of terrestrial vertebrates, up to 96 percent of marine species, and eight insect orders were wiped out.[30] Accordingly, this novel respiration mechanism functioned as a pre-adaptation in diapsids (who gave rise to archosaurs) prior to the mass extinction and allowed archosaurs to thrive in post-extinction low oxygen levels and maintain large body sizes. In contrast, the broncho-alveolar lung, "with its requirement for a thick blood-gas barrier and therefore its intrinsic limitations in low-oxygen environments," constrained mammalian Mesozoic body sizes.[31] Present-day crocodiles have another useful respiratory feature. They can stay underwater for up to two hours because as ectotherms they do not have to produce heat but take it in from their environment. This translates to lower metabolic rates and less energy expenditure, attributes that are very important for large-bodied hunters.

Modern-day crocodiles have five basic styles of locomotion: the high walk, crawling or sprawling, galloping, jumping, and swimming. While many paleontologists assert that the sprawl crawl that a crocodile does when moving along her belly to the water is an overprint of the evolutionary past when archosaurs transitioned to aquatic life, there is growing evidence that crocodilian ancestors were warm-blooded—and bipedal.[32] The faster-moving high walk when a crocodile is fully erect on all four legs reflects these origins. Young crocodiles actually run almost mammalian-like when galloping along at a fast pace. Both the gallop and the high walk are employed to get somewhere on land fast, whether it is to attack prey or escape. Jumping is the near-vertical shooting up out of the water for grabbing prey. And swimming is accomplished with an elegant undulating tail as propeller.

By definition, the apex predator crocodile who can coordinate all of these varied and powerful skills has high marks in what Stanford psychologist Albert Bandura describes as "strong self-efficacy," that is, these crocodiles have "high assurance in their capabilities" and "approach difficult tasks as challenges to be mastered rather than as threats to be avoided." To calibrate the definition of self-efficacy to those not living in the buffered affluence of modernity, we might modify Bandura's definition to read, "high assurance in their capabilities" and "approach difficult tasks as challenges to be mastered *once risks and threats are assessed and can be avoided.*" Bandura goes on to explain that "such an efficacious outlook fosters intrinsic interest and deep engrossment in activities. They set themselves challenging goals and maintain strong commitment to them."[33] Tackling a tapir, even an unmoving body—particularly if his killer is still lurking—is certainly a challenging goal.

An attack, whether to retrieve or kill, once launched, takes "strong commitment." African missionaries saw a full-grown bull giraffe, many who reach twenty feet in height and weigh up to 2,500 pounds, "attempting to negotiate the Olifants River" when he "suddenly stumbled, fell, and was pulled under by a huge crocodile." Elsewhere, an "adult bull cape buffalo (who can reach 11 feet at the withers and weigh a ton) was seized at Nyavutsi waterhole by a 14-foot crocodile and drowned after a tremendous struggle."[34] Crocodiles are able to wrestle and overcome formidable opponents not only because of their size and bite force strength, but also because of their acumen. There is clearly more brain than brawn at work. Vladimir Dinets and colleagues documented mugger crocodiles and American alligators using twigs and branches placed on their snouts to lure nesting birds: the first known example of reptiles using a tool while being mindful of the seasonal habits of their prey.[35]

So far, our psychological focus has been the mindset of crocodiles as hunters. This is natural, because their hunting lifestyles are what grant them membership in the predator and carnivore groups. But there is much more to the poetry-in-motion minds of the Morelet's croc and its kin than procuring a meal. The ecological and physiological complexity that is reflected in the cascade of Latin classifications adhered to every species is mirrored by inner emotional and other psychological complexities. Unsurprisingly then, like the white sharks whom Fred Buyle, Marie Levine, and Ralph Collier describe, every crocodilian is unique, and every individual has a different personality.

Bob Freer moved to Florida in 1970. He had visited the state regularly since 1956, and most of that time was spent with the American alligator and crocodiles. He and his wife, Barbara, founded the Everglades Outpost Wildlife Rescue center in 1991 with the intent to take in injured wildlife, rehabilitate, and release them back to the wild. Bob has handled and moved literally hundreds of alligators labeled "nuisance" animals, "confiscated from illegal or abusive situations by Wildlife, Fish and Game Officers," or "abandoned."[36]

Bob raised from eggs two brother alligators, Lazy and Godzilla, who have been with him now twenty-eight years. When asked about individual alligator personalities and emotions, he said: "Oh boy, are they emotional and very, very smart. Lazy actually lived in the house that was adjacent next to the pond, and he would go in and out when he wanted. After a swim, he would usually come right back in the house and lie down by my feet next to my chair. He'd stay there as long as I would. Then, there is Godzilla who has a completely different personality than Mr. Laid Back Lazy. Even when he was a hatchling, he would charge up to me and a nip my pant leg. He does not like to be around me and doesn't get along with any alligator males either, so he lives with females."[37]

Even the infamously human-hungry saltwater crocodiles show tremendous variation in personality and sociality.[38] Because emotions and sociality are typically associated, intraspecific reptile relationships have been overlooked. But crocodilians are not just workaholics—they like to play.[39] Play ethologists typically use the criteria of homeothermy, a high metabolic rate, extensive parental care, and large brains as requisite qualifications for having fun. But in addition to the growing number of observational studies, the fact that brain structures and processes are shared among reptiles and mammals speaks to an understanding that reptiles also have a *penchant* to play. Psychologist Gordon Burghardt has studied extensively reptilian play and other facets of iguana, turtle, lizard, and snake life experiences. He defines play as "repeated incompletely functional behavior differing from more fully functional versions structurally, contextually, or ontogenetically, and initiated voluntarily when the animal is in a relaxed or unstressed setting."[40] Animal play is an interesting subject, because it expands the usual workaday descriptions of nonhumans as eating, mating, sleeping, and defecating machines. Tossing around a dead gazelle or hippo or enjoying the sensuality of a stream of water doesn't fall into any of the four ethological categories—play is decidedly indulgent, that is, pleasurable with a flair for the fantastic.

In Costa Rica, the friendship of Gilberto (Chito) Sheddon and Pocho, a seventeen-foot, 980-pound crocodile, made global headlines.[41] After a cattle farmer shot the huge crocodile in the left eye, Sheddon hoisted the dying crocodile into his boat and took him home to nurse him back to health. There he cared for the suffering reptile, whom he named Pocho, giving him food and sleeping every night by his side. When Pocho was able to move on his own, the two took to swimming together in the river and playing. As one journalist observed, "Play behavior included swimming together, rushing at Sheddon with an open mouth in mock charges, sneaking up on him from behind as if to startle him, and accepting being caressed, hugged, rotated in the water and kissed on the snout."[42]

At day's end, Pocho would haul himself out of the water and faithfully follow his human friend home. Twenty years later, Sheddon laid his reptilian companion to rest during a public funeral. Some saw the relationship as a commercial ploy, but few could argue that the man wasn't deeply grieved. When asked if he would replace the crocodile with another, Sheddon shook his head, and said "Pocho is Pocho, the only one."[43]

The story of Pocho and Chito's relationship shows just how resistant most scientists are to the idea that reptiles have feelings. Although "Pocho's behavior was seen by thousands of tourists, filmed countless times (including a full-length documentary *The Man Who Swims with Crocodiles* by Roger Horix), and featured by most Central American newspapers," so far, it is maintained, the tale of Pocho has never been mentioned, or if so, minimally, in scientific literature; the only published source of detailed information is a Wikipedia article.[44]

Basking Alligator. Photo credit: Dr. Marge Peppercorn.

Until very recently, reptile sociality, either across or within species, has been ignored or dismissed to the point of illogic. For example, researchers investigating social learning among red-footed tortoises (*Geochelone carbonaria*) claimed that their study was the first to show social learning in a nonsocial species, which, using a kind of *Through the Looking Glass* logic, they suggested that their evidence invalidated the long-held belief that social living precedes social learning. It remains a puzzle as to how and why one would and could learn from fellow reptiles in the absence of having some kind of social organization and relationships.[45] Those more familiar with reptiles and who have been studying how snakes, turtles, lizards, and crocodiles behave with each other pointed out that the "discovery" was not valid because the tortoises are indeed social, very social. Just because reptiles may not live up to mammalians' standards of emotional displays does not mean that they don't have social tendencies.[46] There is even fossil evidence for sociality in some species of dinosaurs.[47]

Aside from well-publicized social species such as African lions and gray wolves, carnivorous hunters are assumed to be independents, preferring to track and catch prey on their own. It has been assumed that crocodiles seem to work the waters solo, but open-minded and interested researchers have been rewarded with the discovery of the social side of crocodilian hunting. Vladimir Dinets observed crocodiles swimming in a circle around a shoal of fish, gradually making the circle tighter until the fish were forced into a tight bait ball. Then the crocodiles took turns cutting across the center of the circle, snatching fish, a strategy reminiscent of certain whales.

Sixty alligators at the Okefenokee National Wildlife Refuge in Georgia tried a different but equally successful approach. They used two coordinated, alternating methods, a driving phalanx of alligators that "would ease itself into a loose semi-circle and then close in, pushing the fish into the shallows and against the bulkheads."[48] Yet another game plan executed by the usually unaccompanied saltwater crocodile entailed one scaring "a pig into running off a trail and into a lagoon where two smaller crocodiles were waiting in ambush—the circumstances suggested that the three crocodiles had anticipated each other's positions and actions without being able to see each other."[49] In Jamaica, although macho tempers may flare at times, there is a palpable social connection among male American crocodiles.

Lawrence Henriques is a biochemist turned polymer engineer who has spent decades studying and caring for these reptiles. He lives on the north coast of his native Jamaica, where he shares his small property and cottage with thirty-two American crocodiles. The sanctuary takes in and cares for rescued, or progeny of rescued, crocodiles. His life's work is to save the reptile from extirpation and restore the species' ability to coexist peacefully and healthily with humans.

Henriques's experience with crocodiles has given him an insider's view. At every moment, he is either walking through the wetlands, swimming in the waters, wading by the shores of crocodile habitat, or sleeping next to the crocodiles. The time spent with the large reptiles is intimate. When they arrive at the sanctuary, they are most often in poor shape—hungry, injured, and dehydrated. Henriques tenderly ministers to elegant webbed hands and feet and damaged scales. The atmosphere is relaxed and open. Sanctuary residents surround his house and may come and go as they please from enclosure pools to the interior of the home. Henriques describes the male ritual at his sanctuary. "Just as dawn breaks, between around 4 and 5 a.m., one of the male crocodiles will slap his tail. Bang. Then, the others follow suit, one after the other. It sounds like a volley of shots going off. It is amazing. The sky explodes with their morning salutation."[50]

Consistent with discoveries in the neurosciences and in developmental biology that have dispelled the erroneous assumption that egg-laying, precocial species lack the affective affinities characteristic of warm and fuzzy altricial species, crocodilian discourse and socioaffective attachment begin well before birth. Crocodile embryos vocalize in their eggs to "fine-tune hatching synchrony and also stimulate mothers to open the nest to free [the] hatchlings and carry them to water." Parental care both before and after the infant reptiles hatch is common.[51] In fact, "crocodilians display a remarkable cascade of maternal care behavior for their eggs and hatchlings, including vocal communication, excavation, carrying hatchlings and eggs to water from the nest, breaking of the eggshell to facilitate hatching, and feeding and protection of the crèche."[52] These behaviors are evidence that crocodiles engage in the same kind of parent-infant neuroendocrinal and psychobiological tuning that has been shown to be so critical in mammalian brain-mind development. In the case of Komodo lizards, who create baby dragons both sexually and parthenogenetically, it is difficult to predict when attachment processes begin and stop.[53]

So, contrary to popular beliefs, we have learned that crocodiles not only are hunters with scalpel-sharp intellects, but also possess and express deep and diverse feelings and a finely tuned ability to use measured restraint and parsimony. Just as connections are forged, however, they can also be torn asunder through stress and violence.

Alligator and crocodile farms are longstanding businesses in the southeast United States. They raise and kill the reptiles to sell for meat, leather goods, and other products. Elephant, snake, monkey, and other exotic and endangered animals have become big, and not always undercover, businesses. Similar to the reproductive cycles of humans in prison and of elephant females in zoos, those of alligators kept in poor conditions and subjected to chronic stress become irregular or "flatline." As Bob Freer has observed: "I've seen at farms where alligators are interfered with a lot and get bothered and harassed, female alligators just dropping their eggs without even burying them. They tend to be highly infertile as well because they are really, really stressed."[54] The crocodile meat trade causes its own stress in free living populations. Lawrence Henriques describes the booming wild crocodile trade in Jamaica that is devastating the already faltering population. Today crocodile flesh is not just readily available, cheap meat but also a trendy menu item among the glitterati.[55] "Our problem here is that we are losing most of our wild crocs to continued habitat loss and more recently, a boom in illicit killing to stock the lucrative meat market. In addition to wealthy Jamaicans, there are a lot of workers from China who enjoy the taste of the exotic and are more than happy to pay big money for a chunk of croc. The crocodiles barely have room to live. Habitat destruction combined with killing and tormenting puts them all under a lot of stress. Most times that I am called to pick up a 'nuisance' croc, he or she is dead or almost dead having been stoned and beaten by people."[56]

In light of what we have learned from bears, elephants, and orcas about attachment and the consequences of relational trauma (that is, maternal indifference, abuse, neglect) and its intergenerational transmission, alligator farms and human impacts on free living crocodilians are breeding generations of compromised and dysregulated individuals. As Bowlby, Schore, Narvaez, and others predict, and which has been extensively documented, such a succession of traumas leads to an array of physiological and psychological disorders that undermine individual, community, and species well-being. Conservation may rely on population numbers and habitat quality to assess species viability, but unless we take action to prevent psychosocial trauma, no amount of preservation will save crocodilians, other reptiles, and other taxa.[57]

It is not as if crocodilians are unaware of what is happening. Similar to the empathic grizzlies who are exquisitely sensitive to shifts in the attitudes of their human ranching neighbors, alligators are better at reading our intentions than we are at knowing theirs. As Henriques puts it, "Alligators and crocodiles are quick studies when it comes to figuring out those who prey on them."[58] Rob Denkhaus, urban wildlife specialist and Natural Resource Manager at the Fort Worth, Texas, Nature Center and Refuge, has focused on the interface between human society and psyche and those of alligators. His goal has been to get people to live cooperatively with, not against, wildlife. In his experience, alligators show every bit of the keen mental acuity and emotional intelligence implied by reptile neuroanatomy:

> I work at a 3,600-acre nature refuge within the city of Fort Worth. Historically, we are at the far northwest range of alligators in the U.S. Since 1913, Texas started putting in surface water reservoirs all over. There are no natural lakes in the state. The dams cut off natural alligator movements. When I came here in 1997, it took me three years to find a gator. Their population throughout their range had been decimated until they were declared endangered in 1973. The law plus education has changed people's attitudes about alligators, and alligators have picked up [on] the shift. Little by little, they've started showing up at the refuge. Now I can with pretty good accuracy tell someone when and where to view one. The alligators are not habituated, not in the way it is defined in the wildlife control world. Alligators are very smart. You're not dealing with a "dumb animal" but someone who is more than on your par. I think they are showing themselves because they know humans have changed—at least within the confines of the refuge. Before, fishermen would always tote a shotgun and if they saw a gator would take a pot shot and so the gators learned to literally lay low and stay out of sight. I'll tell you another thing. Over the eighteen years I've been at the refuge, I have never heard once an alligator bellow. Males make this sound. I think that they learned that bellowing caught humans' attention and their guns so they learned to keep quiet. Maybe it is even a cultural thing now.[59]

After a moment's pause, Denkhaus adds, "I love alligators." That is what Lawrence Henriques says about American crocodiles: "I love crocodiles. That is me in a nutshell. I am Jamaican and I love crocs. They are very intelligent with a propensity to learn. When they are relaxed, when they *can* relax, they are full of joy. That is a normal crocodile. Of course, they each have their personalities and they are each unique, but overall, they are appreciative and pleasant tempered." Henriques continued:

> I'll give you a specific example. I knew a pair of crocodiles who I saw quite often over a five-year period as they lived in a small river near my study area on the southeast corner of the island, close to a culvert. The male was eleven feet and the female eight feet in length. One night on the way to doing fieldwork further east, I only saw her, no sign of him, which was very strange. I searched the creek and found the dead body of the male upside down on his back. Someone had shot him through the head and just left him there. I summoned my police friends from the district nearby to help me recover the body for disposal. With all the activity taking place the female disappeared and it took me a month to eventually find her. She had retreated to a location in an area of mangrove swamp some distance away from where she used to live. I brought her home because I didn't want her to end up like her partner. Doris (I always name the crocodiles who come to live with us and always address them directly by their name when I am around them) was emaciated and traumatized as she was right there when her mate was shot. They were a very close pair. Unlike my other large females at the sanctuary, she is a bit of a bully. Doris won't fight but she displaces. She has settled in but will probably always be affected by the terrible trauma. She has had one set of eggs after mating with one of my young adult males named George but they do not stay together . . . they have not really bonded. Hopefully their young will be able to live in the wild and have a safer life than Doris and her husband did.[60]

Like Fred Buyle who insists that free-diving with sharks and whales does not require any special magic, Henriques, also extremely experienced with the species with whom he works, insists that getting along with crocodiles just takes common sense:

> People don't realize that crocodiles are so very sensitive. When you catch them or move them, you have to be as careful as possible, especially large crocodiles who stress very easily. They can go into shock. According to some reports up to 40 percent die within a week or two after translocation. They don't want to move and often they return to their homes, which usually ends up badly for the croc. Most capture techniques are dangerous and damaging, and prolonged struggling of the animal during capture can lead to severe stress from which the reptile does not recover. I use snares or rope to get them out of the water, restrain quickly and cover the eyes with cloth or tape to reduce stress and keep them cool. No jumping on them as that can rupture their organs. I prefer to do all of this myself. If it is data collection, no problem; but if the reptile is to be moved, a few extra hands are sometimes required. If the corpse of a dead reptile is found, whether it is in the field or while in captivity, it is removed at night so as to minimize the exposure of this unfortunate activity to other reptiles, who may be curious.[61]

One of the residents at his sanctuary is a thirteen-foot, six-hundred-plus-pound male named Brutus. Two years ago, Henriques got a call from the police saying that an adult male crocodile was sitting on top of a rubbish heap, broken glass and plastic containers all around, scavenging: "I got there at 11 that night. I had no problem restraining him, but being so big[, he] required the combined efforts of ten police, one kind truck driver, and myself to get him into the Land Rover, still leaving part of his tail hanging out of the back, when I drove off. Brutus is a big baby—a lot of adult male American crocodiles are. So good-natured, and he is a great ambassador. They all are, those who, unlike Doris, did not have such a nasty shock and raw deal."[62]

These personal tragedies provide a face for the mind-boggling global decimation of reptiles. A 2013 comprehensive report by more than two hundred authors on the global status of reptiles found that "nearly one in five reptilian species are threatened with extinction, with another one in five species classed as Data Deficient," meaning the news is probably worse because absence of data implies understudied and overlooked.[63] The "Data Deficient" species are subject to the same kinds of stressors and damage as any others. Almost all populations of the crocodilian order have plummeted and are still nose-diving because of exploitation, persistent legal and illegal hunting, and habitat destruction. Along with its compatriots, the King cobra and the Indian gharial, the mugger is locally extinct in many areas including Bhutan and Myanmar.[64]

Human persecution of crocodiles is deeply rooted, beginning with European occupation of North America, Asia, and Africa. In a historical review of attitudes toward the Nile crocodile, Zambian Mwelwa Musambachime quotes diverse sources depicting European hatred of the reptile: "H. L. Duff, employed by the Nyasaland Administration in the early 1900s, believed that the crocodiles were 'to be shot whenever and wherever they were seen' . . . it was 'nothing short' of his duty 'to mercilessly destroy [every crocodile] which gave him the opportunity' . . . Muirhead, a self-confessed poacher[,] proclaimed that, in shooting crocodiles, he was not 'robbing Africa of something that will never return but ridding it of a pest' . . . Holmes, Barns and Gregory found shooting crocodiles a very 'good past-time' . . . Pitman, the Chief Game Warden[,] regarded the crocodile as a 'foul beast' and for 17 years encouraged its extermination. In the eyes of many Europeans, the crocodile stood condemned and had to be eradicated from African rivers at all costs."[65]

Human fear and hubris were driving factors, but huge economic interests were also a motivation. Rivers and lakes were cleared of crocodiles to boost commercial fishing and to provide a lucrative sport for big game hunters (along with the usual list of money-making and power-acquisition agendas that defined colonialism). Crocodiles and alligators have been mercilessly hunted for their skins and flesh. The golden scales upon which crocodiles "pour the waters of the Nile" were a mainstay of the fashionable in shoes, pocket books, and various and sundry other items as well as haute and not so haute cuisine.[66] The attitude that the crocodile was a scourge even permeated science.

Scientist Alistair Graham and photographer Peter Beard, who at the time

The social Crocodile. Photo credit: Lawrence Henriques.

were both regarded as staunch conservationists, coauthored *Eyelids of Morning,* which chronicles in rich text and photos their 1966 study of the Nile crocodile in Lake Rudolf. In the name of science, their expedition killed more than five hundred crocodiles out of a total of 14,000 (almost 4 percent). Many were huge, older individuals. The majority of photos in the book show dead crocodiles in diverse poses. Perhaps mindful of a changing social political climate, in the preface to the book Graham includes this disclaimer: "To many people, the pictures of dissected crocodiles will appear as so many repellent 'death images' and our activities altogether dubious. But scientists cannot let susceptibility to imagery or sentiment cloud their vision, for if they do they will cease to make impartial observation[s]."[67]

The authors go on to offer what seems an anticipatory defensive apologia for the slaughter: "For the most part, man gets on with crocodiles about as well as he did with dragons; and just as he did with dragons, he will banish them from all the remotest parts of the earth . . . for civilised man will not tolerate wild beasts that eat his children, his cattle, or even fish he deems to be his; that would be regression into barbarism."[68]

As biologist Joe Wasilewski describes, our scaly friends are burdened with bad press and "sensationalism," as well as a public that expends little effort at finding out who is really behind those bared teeth and fangs. Buyle concurs, shaking his head:

> It is sad. Crocodiles have the same problem that sharks do. You realize that so much of the problems that these carnivorous species have is caused by an image problem. They are persecuted, tortured and killed by the millions merely because of human prejudice, our insistence that someone has to look like us to be accepted. Sharks are the "bad guys" because they don't look like a mammal, they look like fish and they have these seeming emotionless faces and big nasty teeth. On the other hand, it is a completely different story with dolphins. They are the quintessential "good guys" because, well, Flipper was a nice fellow and dolphins smile all the time. It is not a smile of course, but people don't bother to learn about that, and the fact that dolphins do a lot of not-nice things—like raping females and killing baby dolphins. I guess that is the bottom line: humans judge someone by how much the other person is like or different from them. If they are the same or similar, then that is okay. If not, well, too bad for you . . . Life would be a lot nicer, and the world a lot better, if people didn't judge others from the outside alone.[69]

But as in the case of grizzlies, until white settlers and colonists took over the planet, most people learned how to live with species that might eat them. In many, if not most, areas, crocodiles and humans lived in what Henriques calls "commensal respect." W. H. Bennett, a missionary who spent years in Zaire, wrote that the communities believed that "the crocodile of itself is a harmless creature and [so] thoroughly do they believe this that in some places they go into the river to catch fish with a bait and attend to their traps without hesitation." Europeans were "surprised at the indifference of the African people to crocodiles so that often they crossed rivers even when neck-deep without hesitation and others went into the rivers and lakes to bathe, wash their clothes and draw water without taking the slightest precaution at all. The belief among these people was that crocodile attacks were the work of witches or were a form of punishment for wrong-doing meted out to the victims."

Colonist perceptions and understanding of "native indifference" was often rather simplistic and overlooked deeper cultural and spiritual underpinnings, as a report by James Brooke shows. In the early nineteenth century, Brooke ruled as Rajah of Sarawek, Borneo, home to the legendary *Bujang Senang,* white crocodile. He recalled how once he encountered a man whose leg had been mutilated by what he calls an "alligator" but was probably a saltwater crocodile who inhabited the region: "I asked him if he had since retaliated on the alligator tribe. He replied 'No; I never wish to kill an alligator, as the dreams of my forefathers have always forbidden such acts; and I can't tell why an alligator should have attacked me, unless he mistook me for a stranger, and that was the reason the spirits saved my life.'"[70] This belief persists today, although Westerners and Western-inculcated Borneans, eager to increase river exploitation of prawns, have tried to retaliate by trying to get crocodiles delisted.[71]

Not all indigenous cultures have reacted this way, however. The Tana-wa-Pokomo who lived on the shores of Lake Turkana, Northern Kenya, "regarded crocodiles as their 'most deadly foes' and waged a war of extermination against them."[72] The apparent correlation between human aggression and crocodile aggression evokes the old adage "violence begets violence," however, Lawrence Henriques and others assert that alligators and crocodiles become more shy and retiring than aggressive when met with hostility. Although it seems that respectful co-existence prevailed in most locales, Musambachime describes how the

> behaviour of [Nile] crocodiles towards man varied from place to place. In some locations, such as on the shores of Lake Mweru, and along the Shire and Lualaba rivers, crocodiles were very aggressive and dangerous so that palisades and stockades had to be

> built on the edge of the river bank or lake shore where people bathed or drew water. In other places, contraptions similar to the *shaduf* used in Sudan and Egypt were constructed above the shore or river bank to allow women and children to draw water without risking being caught by crocodiles . . . In other areas, by contrast, for example along parts of the Congo (now Zaire) River, no precautions were taken whatsoever and few injuries were sustained . . . So it seems likely that the aggression in the crocodile might be due not to some naturalistic characteristic but to the non-availability of alternative prey.[73]

A further taxonomic note: Congo crocodiles are now believed to be a different, smaller, less "dangerous" species.

Not all crocodiles attack humans, not all attacks are fatal or unprovoked, and the number of bites and deaths are vanishingly small when compared to run-of-the-mill activities of modern life—driving, taking a shower, having surgery—which produce orders of magnitude more human fatalities. Statistics of fatal attacks are difficult to assess by species, because not all attacks are reported with the same frequency across countries of origin. The reasons have to do with differences in human cultural attitudes, what might be called risk perception, and access and motivation for reporting. Long-term trends are not easy to interpret. Statistics are confounded by several factors that involve both crocodilians and humans, including people's varied readiness to report an incident; species identifications that are often of dubious accuracy; the frequency of human incursions into crocodile habitats; and availability of food and high-quality habitat for crocodiles. Subsequently, we have only rough estimates on the numbers and reasons for these attacks.[74]

Generally, attack frequency on humans correlates with body and mouth size.[75] A croc's eyes may be bigger than her stomach, but like sharks, she usually doesn't go after anyone much bigger than she can handle. For example, young Nile crocodiles living in Africa's Okavango Delta start off modestly eating insects and crustaceans until their heads and mouths broaden sufficiently to take on the larger prey that frequent the shores of streams and lakes.[76] Australian crocodile researcher Matthew Brien has commented, "A two-metre croc won't eat a human, but once you get an animal that is four metres long, humans are certainly on the menu." He then gives some sage advice: "You really do have to take precautions in croc habitat."[77] An obvious warning that tourists and resort businesses should heed in regions where the toothed reptiles reside. A toddler who was visiting Disney World with his parents in 2016 was attacked and killed by an alligator. Although the boy's body was found elsewhere, four alligators were summarily executed and cut open.[78]

A more accurate predictor of carnivore prey selection is bite force.[79] To be able to exert considerable force, one's skull must be up to the task. This is not the case for the Komodo. Relative to the formidable saltwater crocodile, the Komodo skull is quite fragile and incapable of inflicting a powerful bite. Instead, researchers posit that the Komodo uses a "sophisticated combined-arsenal" to fell their prey. With their razor sharp teeth and skill at landing an accurate slashing cut, the accompanying Komodo venom causes anticoagulation and shock induction.[80] This theory is buttressed by the fact that pathogenic bacteria, formerly regarded as the lethal agent, were not found in the mouths of free-living individual Komodos when tested.

Beyond the blur of statistics and circumstance, even though many mature crocodiles are certainly capable of attacking and deftly killing humans, they rarely do. It is generally accepted that most attacks occur as a result of human provocation or a too-good-to-pass-up opportunity. As Henriques says, "Humans are easy."[81] This is particularly true when someone is less than fully equipped, as was the case of an inebriated couple who went swimming in a Florida canal, and it is reported, "messed with" an American crocodile at 2 a.m.[82] They were bitten but not pulled underwater, nor did the crocodile try to eat them. This is the first crocodile attack reported in the United States. Most cognoscenti agree, however, that there are two species who gladly and routinely snack on humans when given the chance: the Nile and saltwater. (The majority of these attacks are by males of the species.)[83] According to the IUCN-SSC Crocodile Specialist Group, "there is little doubt that the Nile Crocodile is responsible for more attacks on humans than any other crocodilian species. After lions and hippos, the Nile Crocodile causes the highest numbers of wildlife-related fatalities in Africa . . . [exceeding]

300 per year . . . The number of attacks by saltwater Crocodiles, throughout the species' range, is estimated to be around 20–30 per year."[84]

Why is it that some species get big enough to eat humans, but they do not, and others, such as the Nile and saltwater crocodiles, seem to have a strong predilection to eat our species? Although others tended to be puzzled or attribute the seemingly preemptive aggression to an innately bad temper, Henriques offers this reply:

> It is the difference in the ways that the two species evolved. The Nile and American crocodile came from the same Pangean ancestor so the two are related. But the size of prey stock in Central American and Caribbean waters are small compared to, say, Africa. American crocodiles mainly eat fish and crabs, and the occasional small mammal and birds. It may be because African crocs have coevolved with big game. Their bodies are larger, bulkier, and their heads are robust. You need all this "gear" when you have to take down a six-foot sharp-hooved wildebeest or eight-hundred-pound kicking zebra. If you are going to eat a big animal with horns, then you have to be aggressive. You have to punch hard. One slicing contact with a hoof and the croc is dead. Despite their hard scales, they are very vulnerable to something like a panicked hoof[ed animal] fighting for his life. So attacks have to be decisive, no shilly shallying. You have to be strong, committed, and smart. Furthermore, African crocs don't often get to feed regularly. When a migration comes, the rains come—then bam! Time to eat and feast. Then nothing maybe for up to six months.[85]

Perhaps this is "where all the aggression comes from and why humans are an appealing meal for these big crocodiles."[86] And big meals are key for survival. Whatever the explanation, we witness once again, as in the case of orcas and sharks, sensitivity and a broad range of cultural values and traditions among the crocodilians.

But it does seem that the need for food drives attacks on humans. Today's drive-in, takeout urban dwellers forget that finding and securing food is

Lawrence Henriques with Brutus at their Jamaican sanctuary. Photo credit: Lawrence Henriques.

paramount. Australian philosopher Val Plumwood realized this hard truth when her canoe was overturned by a saltwater crocodile while paddling in the East Alligator River of Kakadu's paperbark wetlands: "The water lilies and the wonderful bird life had enticed me into a joyous afternoon's idyll as I ventured onto the East Alligator Lagoon for the first time in a canoe lent by the park service . . . [I] glutted myself on the magical beauty and bird life of the lily lagoons, untroubled by crocodiles . . . I set off on a day trip in search of an Aboriginal rock art site across the lagoon and up a side channel."

Then the day's idyll vanished. The sun turned to drizzle and drizzle turned to wind and rain. Plumwood found herself lost in a maze of backwater channels. Suddenly she realized she had entered the main river where the park ranger had specifically warned her to avoid on account of crocodiles. Plumwood was not new to the outback. She considered that she possessed "considerable bush experience," so she kept watch. "As I looked, my whispering sense of unease turned into a shout of danger . . . As a solitary specimen of a major prey species of the saltwater crocodile, I was standing in one of the most dangerous places on earth . . . I had not gone more than five or ten minutes down the channel when, rounding a bend, I saw in midstream what looked like a floating stick, one I did not recall passing on my way up. As the current moved me toward it, the stick developed eyes. A crocodile!"

Her fears materialized:

> I was totally unprepared for the great blow when it struck the canoe. Again it struck, again and again, now from behind, shuddering the flimsy craft. As I paddled furiously, the blows continued. The unheard of was happening; the canoe was under attack! . . I steered to the tree and stood up to jump. At the same instant, the crocodile rushed up alongside the canoe, and its beautiful, flecked golden eyes looked straight into mine . . . Before my foot even tripped the first branch, I had a blurred, incredulous vision of great toothed jaws bursting from the water. Then I was seized between the legs in a red-hot pincer grip and whirled into the suffocating wet darkness . . . [the death roll] is . . . an experience beyond words of total terror . . . The roll was a centrifuge of boiling blackness that lasted for an eternity . . . [then] the rolling suddenly stopped . . . The crocodile still had me in its pincer grip between the legs. I had just begun to weep for the prospects of my mangled body when the crocodile pitched me suddenly into a second death roll . . . the crocodile seized me again, this time around the upper left thigh, and pulled me under.[87]

Remarkably, Plumwood managed to escape the crocodile's grip and pull herself up the bank, crawling to safety. She was bleeding profusely from the gash in her leg, which, without the lucky intervention of the concerned park ranger who went in search of her, would in all likelihood have proven fatal.

Those familiar with salties are amazed at her escape. The saltwater crocodile is described in the hushed tone of awe as "a formidable predator."[88] Males commonly reach six meters in length and up to a ton in body mass. Plumwood had survived not one but three death rolls. A crocodilian death roll has nothing to do with sushi, although the recipient might feel like it when locked between the jaws of a giant reptile set on acquiring a full stomach. The spiraling roll, where the victim is clasped underwater in the clawed arms and teeth of a crocodile, is a technique used to drown and dismember large prey.

In the wake of her recovery, and before her death nearly twenty years later, Plumwood reflected on the profound experience that she had had: "I leapt through the eye of the crocodile into what seemed also a *parallel universe,* one with completely different rules to the 'normal universe.' That is something that sends you spinning through the tunnel at a frightening speed to arrive with a very big bump, a terrible bump, in the parallel universe. For in the death moment we are reclaimed once again by the sphere of nature, of death as nature, and especially so when we die by a predator of nature—as do the other beings we have said are not of our sphere."[89]

Within suspended near-death moments, the conventional image of us (humans) versus them (carnivores) that has been archetypically, if not genetically, engrained in our species was dislodged for Plumwood, who found "the world I thought I lived in to be illusory, my own view of it terribly, shockingly mistaken at the moment of being grabbed by those powerful jaws, that there was something profoundly and incredibly wrong in what was happening, *some sort of mistaken identity.* My disbelief was not just existential but ethical—this wasn't happening, couldn't be happening . . . The creature was breaking the rules, was totally mistaken, utterly wrong to think I could be reduced to food. As a human being, I was so *much more than food.*"[90] This is exactly what science has done. By demonstrating that reptiles possess the qualities thought to be uniquely human, neurobiologists have shown that *crocodiles and alligators are also more than food* and much more closely related to our own species than what most people imagine. As Plumwood reflects, "I knew I was food for crocodiles, that my body, like theirs, was made of meat. But then again in some very important way, I did *not* know it, absolutely rejected it. Somehow, the fact of being food for others had not seemed real, not in the way it did now, as I stood in my canoe in the beating rain staring down into the beautiful, gold-flecked eyes of the crocodile. Until that moment, I knew that I was food in the same remote, abstract way that I knew I was animal, was mortal. In the moment of truth, abstract knowledge becomes concrete."[91]

Theoretically, according to comparative and evolutionary neuroscience, each member in the family tree carries with it the thoughts and learning of its ancestors—an idea that is embodied in C. G. Jung's schema of consciousness, which describes the ancestral collective lurking in the depths of the unconscious while the personal consciousness bobs in the waters of the individual present. Despite our species' insistence that we are separate from the rest of nature, the memories and motivations of our phylogenetic heritage remain in our neurobiological bones. Crocodiles and humans likely share archetypes.

Although the foraging techniques of crocodiles and our own ancestors may have diverged from those of humans and crocodiles today, the same kinds of relational brain functions are at work. Think back to the moments when you read the river scene in which the huge reptile lay submerged, waiting, as they say, for the kill. Did the hair on the back of your neck lift ever so slightly? If so, you have an age-old appreciation of someone very smart and perhaps smarter than you. This someone has to be able to navigate land and water, hold her breath, and move along the river bottom only to rise like a submarine at exactly the right spot to spy or launch an attack. It is this felt connection, a spark of recognition, that points to our overlapping evolution and relatedness to reptiles.

Even after all his years of intimacy with reptiles, Lawrence Henriques maintains that "we haven't scratched the surface of crocodile minds."[92] He continues his gentle care and restoration work with the American crocodiles in Jamaica—showing by example a measured, informed, and respectful way of living with animals who, as neuroscience shows, have demonstrated much greater emotional intelligence than our own species.

5

Rattlesnakes: Sensibilities and Social Life

Cicadas are perfect thermometers. When 64° Fahrenheit is reached, they begin to sing and every inch of air is infused with their melody. These insects are some of the brave few who show themselves midday in this new global-warming era, but even they are daunted in relentless triple-digit heat. Everyone contracts to a minimum of movement, as if air itself were finite. Ravens retreat to their creekside eyries, deer rest below buckbrush, rattlesnakes seek the cool of a rock's underside, and their nemesis, the ground squirrel, estivates.

Estivate! Biologists have such wonderful words. According to etymologists, the term was coined in 1650 and derives from the Latin *aestivatus,* "to

spend the summer."[1] When the temperatures get too hot and conditions too dry, California ground squirrels (*Otospermophilus beecheyi*) estivate.[2] This retreating phenomenon is similar to hibernation. Metabolic rates and body temperatures decrease during periods of both cold and warm weather dormancies. But estivation occurs in summer and is of shorter duration, whereas hibernation, depending on local parameters, extends for several months during winter.

When not estivating, ground squirrels are unsurpassed by their tenacious chewing and foraging. Nothing and no one stands in their way. They can and will gnaw their way through wood, plastic, and whatever material lies between them and what they want to eat. The intrepid ground squirrel is, however, vulnerable to certain meat-eaters such as coyotes, feral felines—and rattlesnakes. Rattlesnakes are generally regarded as ambush carnivores who lie in wait for delicious prey to appear. Depending on the rattlesnake's size, species, and location, he will eat rabbits, all sorts of rodents, birds, lizards, some snakes, and even insects at times. Although some eschew leftovers, others species are a bit less picky and take what they can.[3] Among the favorite foods of the patterned snakes are young ground squirrels. As it turns out, the rodent-snake relationship is quite complex and provides intriguing insights into rattlesnake psychology.

Rattlesnakes belong to Crotalinae, or pit vipers, a subfamily of Viperdae. "Pit" refers to two tiny concavities located between the eyes and nose of the snake that, coupled with the body's powerful coiled launch, acts like an infrared-sensing missile. Rattlers are not the only species to possess these special thermoreceptors—the common vampire bat (*Desmodus rotundus*) has them as well.[4] These sanguinary gourmands tune their thermoreceptors to detect the temperatures of species with whom they have evolved.[5] It was not until 1935 that the function of the pits was identified. A German ethologist, M. Ros, observed that her pet African rock python (*Python sebae*) failed to gravitate to a warm rock after she applied petroleum jelly to cover his pits.[6]

(Overleaf) Henry, Western Diamondback Rattlesnake. Photo credit: Melissa Amarello and Jeff Smith.

There are interesting aspects to pit neurothermo properties. Unlike other sensory organs, pit infrared receptors are never quiet. Their neurons fire off and on, day and night. Although not everything is moving, every living being gives off heat, which can then be picked up by infrared pits. Spontaneous discharge performs as a kind of ever-alert sonar sweep. This is a critical feature if danger or food is to be identified in a timely manner.

There are about thirty species and several scores of subspecies in the rattle-making genus *Crotalus.* Beyond *Crotalus,* there is a second genus, *Sistrurus,* with only three species—the massasauga (*S. catenatus*), pygmy (*S. miliaris*), and Mexican (*S. ravus*) rattlesnakes—who hail from south and east of the Rockies. As genetic testing methods have become more common and begun to replace climatic classifications, evidence has emerged indicating that some hybridization may occur among species, although rarely.[7] Mitochondrial DNA and ATPase tests taken from the shed skin of a snake with blended markings showed that he or she was a cross between a western diamond-backed rattlesnake (*Crotalus atrox*) and a timber rattlesnake (*Crotalus horridus*).[8]

Although rattlesnake species share certain characteristics, researchers make the point that each has its unique psycho-ecological signatures and habits. As we now know from Bowlby and the bears and neuropsychology, differences in habits imply differences in mind. For example, not all rodents and rattlesnakes live at odds.

Roger Repp of the National Optical Astronomy Observatory and a decades-long researcher of the western diamond-backed rattlesnake (*Crotalus atrox*) describes a cozier version of rodent-rattler relationships than the ground squirrel–rattler pairing: "I have many observations of strange interactions between pack rats [one of multiple species of the genus *Neotoma*] and *atrox* during the colder months. On November 25, 1995, a group of six of us saw a pack rat resting with one of its paws against the flanks of a very large *atrox.* One week later, on a solo jaunt, I saw what appeared to be the same rat snuggled against the flanks of the same snake. The rat was sleeping when first viewed, and only lazily opened one eye when light bouncing off my mirror hit the rat in the face. It does not surprise me that the snakes do

Asa, California Ground Squirrel. Photo credit: G.A. Bradshaw.

not appear to eat the rats during the colder months. What does surprise me is the fact that the rats seem to know they are safe."[9]

Even though there may be convivial reptile-mammal arrangements, rodents remain a culinary staple for rattlesnakes, and it is serious business. Finding and catching a mouse takes more than hanging out a shingle. Snakes do not have to eat every day, but too many missed meals are not advisable. In much the same way a city commuter needs to know the ins and outs of bus schedules and routes, a rattlesnake has to know optimal stakeouts if she is to find food efficiently. Successful hunting entails knowing where mice roam, forage, and live without the huntress being found out herself. All of this takes planning, as Cornell herpetologist Harry Greene saw in action.

One day in Arizona, Greene came upon a longtime friend, a male black-tailed rattler (*Crotalus molossus*), and was treated to a behind-the-scenes glimpse of a rattlesnake trying to solve a dilemma: "No. 41, a large male, crawled into a shady ravine, tongue-flicked around a rodent runway for 13 minutes, and moved back into a hunting coil with his head aimed across the prospective ambush site. Then, after two minutes, he extended his head and neck in a stereotyped posture typically used to fight with other males and pressed down a dried fern a few inches in front of him . . . that might thwart his strike."[10]

Greene later found out that the black-tailed's use of an extended "shepherd's crook" posture to depress plants before retracting into the iconic ambush coil, "otherwise used during male-male combat around nearby receptive females," has been observed in other instances. Prairie rattlesnakes (*Crotalus viridis*) engage in a similar posture to optimize their deer mice (*Peromyscus maniculatus*) hunting success and northern Pacific rattlesnakes (*Crotalus oreganus*) "modify overhanging vegetation to create ambush windows for hunting California ground squirrels."[11] Greene concludes, "Clearly, rattlesnakes can solve barrier problems, and it stretches credulity to not attribute intentionality to their responses."[12] Such careful strategizing is requisite for hunters who execute highly sophisticated surface-to-air strikes, which in turn are informed by their equally sophisticated method of detection. Further, all of this, as we saw in the case of the Morelet's crocodile, requires self-awareness, empathy, and competence in order to integrate, evaluate, and act on information with efficacy.

The snake's facial pit is divided by a gossamer-thin membrane characterized by a high concentration of terminal nerve masses in which differences in external heat—variations as small as 0.001–0.003° Celsius—are detected.[13] This allows the snake to catch the heat signatures of prey. A rattler's infrared receptors absorb signals at two main wavelengths, 3–5 and 8–12 micrometers, which correspond to the thermal emissions of their endothermic prey.[14] Hot blood, hot bodies. The converse happens when a drop in temperature is detected. Because different bodies and materials such as rocks and plants increase or decrease their temperatures at various rates as the sun rises and sets, respectively, snakes are able to discern who and what are around. Pretty neat. But there's more.

In essence, rattlesnakes have four eyes, two normal vertebrate eyes that work in tandem with the other two infrared pits. Rattlers may seem unblinking, but it is because their "ocular scale" or *brille,* the German term for "glasses," makes them appear that way. The brille is a wonderful desert goggle that keeps dirt and sand out. It is transparent and does not move except when it is shed during molting.

The two infrared pits enhance vision by expanding the electromagnetic range over which the snake can gain information about the environment. Infrared and visual information skips in parallel down their respective neuronal pathways to converge and be interpreted by the central nervous system. In this way, pit "infrared receptors are an integral part of the snakes' visual system that makes use of the longer waves of the electromagnetic spectrum for which there are no appropriate photoreceptive pigments in nature."[15] Richard Goris makes the point that "just as the world that most insects see includes both the visual and the ultraviolet spectra, so the world that boas, pythons, and pit vipers see includes both the visual and the infrared spectra."[16] Then, too, rattlesnakes have to deal with the complicated habits and ways of their prey.

Ground squirrels have a highly structured communal life, much of which takes place in networks of tunnels and dens with multiple individualized entrances. If there is one squirrel, there are always more. In this case, many hands do not make the work of foraging any lighter, but they do make protective vigilance more effective. At the first sight, sound, or scent of danger, a squirrel will let off a repeating alarm cry that sends all the squirrels nearby dashing to safety underground, where they will stay hidden until the perceived danger is judged to have passed. Although they use vocalizations to sound off at eagles, coyotes, raccoons, and foxes, ground squirrels use a different repertoire with snakes.

The primary defense strategy used by ground squirrels against snakes is prevention. The squirrel employs "sand kicking": using her feet to spray dirt at the snake for the purpose, biologists surmise, of alerting the snake and causing him to rattle his dry dermal scales. Researchers interpret the squirrel's brazen act as a way of letting the snake know that she is wise to him. When the snake recoils and begins to rattle, the squirrel is able to gauge the size of her opponent and his attitude—that is, his inclination and ability to strike.[17] If the snake is big and hungry, then the ground squirrel may have to go to Plan B. But if the snake is small or is sleeping off a recent meal, then Plan A (scare snake away) will probably work.

Ground squirrels have another trick—they are able to stave off the toxic effects of snake venom. Researchers attribute the squirrels' enhanced immunity to the evolution of venom-neutralizing blood proteins, and/or, some maintain, by their practice of chewing on snakeskin. Cotton rats (*Sigmodon hispidus*), gray wood rats (*Neotoma micropus*), and opossum (*Didelphis virginiana*) also show immunity to the venom of western diamond-backed rattlesnakes.[18] In most instances, an adult won't die from a bite. Sand kicking, immunity to snake venom, and the ability to perform Cirque de Soleil–style acrobatic leaps that propel, twist, and turn the ground squirrel's lithe body are capabilities topped only by ground squirrels' *pièces de résistance,* their tails.[19]

When a ground squirrel encounters a rattlesnake, she will begin swishing her tail in a windshield wiper motion.[20] The move is remarkably successful at getting the snake to back down and switch from predatory to defensive mode. This kind of visual display made sense during daylight hours, but it puzzled researchers when they saw that squirrels employed the method even under conditions of limited light, at dusk when snakes go on the prowl or in a darkened lab room. How, they wondered, could tail swishing be detected when light conditions favored the infrared-sensitive snake and hindered visually oriented squirrels? Could it be that the squirrel was using the same infrared channel as the snake when swishing her tail?

Aaron Rundus, professor of psychology at West Chester University of Pennsylvania, and his colleagues set up a field experiment and found that their supposition was correct. Infrared videography showed that when a squirrel swishes her tail at a rattlesnake, the tail lights up. It seems that ground squirrels use a heat-dissipating mechanism to raise their tail temperatures, related to thermoregulating methods used for hibernation. Furthermore, their tails did *not* light up when the squirrels encountered nonvenomous gopher snakes (*Pituophis melanoleucus*). To the untrained eye, gopher snakes look very similar to certain rattlesnakes and they even engage in similar choreographed coiling and rattling, although no sound emits. Venomless gopher snakes are also fond of eating squirrels, but they lack the infrared-sensitive pits that favor rattlers. To a gopher snake, an infrared-flashing squirrel tail would be meaningless. The motion of the threatening tail is enough to make the gopher snake back off.

In the case of a rattlesnake, though, researchers became convinced that the squirrel turns on her tail-light to ensure the message gets through in rattle-ese. To test this hypothesis, the scientists constructed a "robosquirrel," controlled remotely, that would glide back and forth on a runner and swish its tail. The "Generation One" robosquirrel made for this experiment was computer controlled to remove any experimenter bias and did not move back and forth. Yet Rundus could use it to independently heat the body and the tail regions and to move the "tail flag at a variable rate contingent on the distance of the rattlesnake to the squirrel model."[21] By varying the robosquirrel's actions, the researchers could see how the snake behaved during these encounters. When the robosquirrel's tail swished, the rattlesnake would shift from predatory to defensive mode, but when the robosquirrel did not swish his tail, the snake struck. The major finding, according to Rundus, "was that the visual aspects of tail flagging moved rattlesnakes from predatory to defensive behavior (when tail was not heated) but when the tail was heated the shift was much more pronounced. In fact, the most defensive behavior rattlesnakes exhibit, rattling, was only observed when the tail was heated, not during unheated tail flagging."[22]

Other rodents use thermal communication skills. A similar tail flagging/snake-deterrent behavior has also been observed in chipmunks (*Tamias striatus*) and gray squirrels (*Sciurus carolinensis*), who are able to fend off timber rattlers.[23] Interestingly, in their studies of the Arizona black rattlesnake (*Crotalus cerberus*), Melissa Amarello and Jeff Smith, herpetologists and co-founders of the non-profit Advocates for Snake Preservation, noticed that ground squirrels and other "snake harassers" (as rattler advocates call them) such as chipmunks and rock squirrels (*Otospermophilus variegatus*) seem to initiate contact by coming right up to the snakes' den doors between April and May. That young squirrels do not show up until much later suggests that the adult chipmunks and rock squirrels are making sure that rattlesnakes know not to plan on exploiting their young as easy food. Rundus maintains that, to his knowledge, "thermal signaling by other species has not been demonstrated yet. These species do tail flag but no thermal signaling has been studied, but it is possible that they do."[24] In any case, given the intricacy of the strategies and communication systems that have coevolved with the species, it is clear that they share a set of complex and interrelated behaviors—and minds.

Before leaving the world of squirrels and going on to other aspects of rattlesnake minds and experience, there is one more thing to know about the psychology of the intrepid ground squirrel. One area of scientific interest has been probing when and why a mother squirrel would put her life on the line and intervene with a rattler to try to save her progeny. Researchers tested two ideas. The first was based on the "offspring value hypothesis," which predicts that mother ground squirrels will choose to safeguard older children "because they are more likely to survive to reproductive age." Conversely, the second, "vulnerability hypothesis," predicts that because "older offspring are less vulnerable to predators . . . mothers should take fewer risks" and so preferentially choose to protect younger children. When tested, the mothers' actions did not conform to either hypothesis. They confronted rattlesnakes no matter what their progeny's age.[25] So it seems that ground squirrels don't buy into any age-related model of evolutionary kickback. Instead, their actions appear to be dictated by the bonds and feelings for their young, no matter what age.

By delving into the motivations and mechanisms of communication, research on the interactions of ground squirrels and snakes offers an unusual, but rich, glimpse into the interpersonal psychology of predator and prey. Much of traditional ethology and biology celebrates nature's splendor, but their studies usually stop at the interface of skin and scale and ignore who lies within. For example, a study published in 2015 in *Nature* on the dermal mechanisms of communication by the Madagascar panther chameleon (*Furcifer pardalis*) reports on the squamate reptile's "ability to exhibit complex and rapid colour changes" with comparably mind-bending convolutions. Most of the article is filled with complex technical details describing how the animal's iridophores "share the same neural-crest origin as pigmented chromatophores, the active tuning of guanine crystal spacing we describe here could be considered analogous to movements of pigment-containing organelles in other types of chromatophores . . ." The narrative goes on in even further detail.[26]

But the complicated descriptions, which are certainly fascinating from the perspective of biochemistry, leave the reader a bit awash in reductionistic minutia and wondering what passions lie beneath the chameleon's stupendous color show. For all of its incisive analysis, only a handful of words allude to anything remotely suggestive of the state of mind and inner emotions that must accompany the chameleon's incredible firework display. Given neuroscience's lessons of the crocodile brain, it would be stranger than fiction if chameleons did not also share the socioaffective capacities found in warm-blooded species.

In contrast, reminiscent of John Bowlby, rattlesnake-squirrel infrared research, while also being technically replete, draws attention to the thinking, experiential, and relational, *interpersonal* world of squirrels and snakes. The subject of the study, the rattlesnake, is presented as a psychologically attuned individual who feels (is wary of the squirrel) and thinks (calculates whether to fight or retreat) in empathetic relationship with his prey—a cross-taxonomic class relationship, no less.[27] These insights provide a welcomed portraiture.

Historically, a rattler has been considered to be "a snake without a friend," who by force of circumstance must rely on the kindness of strangers. But there is now a growing cadre of admirers who have done much to change the negative image.[28] Laurence Klauber was one of the earliest investigators of the American pit viper. Although many details have been amended since its first publication, Klauber's 1956 two-volume text is regarded as a classic. A reptile enthusiast turned expert and the first curator of reptiles at the San Diego Zoo, Klauber is regarded as the grandfather of rattlesnake natural history. Klauber recognized how ignorance and myth have distorted our understanding of snakes and created a larger-than-life fear. He recounts how a California man, after being bitten by a look-alike, nonvenomous gopher snake whom he took for a rattler, almost died from shock. "Many persons think that all venomous snakes are aggressive or even vicious and vindictive creatures, whereas harmless snakes are thought to be kindly and timid. So, when they see a harmless garter or bull snake go through all the actions usually attributed to a rattler, such as coiling and striking, flattening the body and hissing, and even vibrating its tail—which when done among dry leaves, makes a fair imitation of a rattle—they jump to the conclusion that the snake is a rattler, or at least a close relative, even though it lacks the telltale appendage on the tail."[29] Much like fish tales, few characteristics "are of greater importance to the average person than [rattlesnake] size. When a rattlesnake story is told, the size of the snake is always an important feature." But, he qualifies, it is not only lurid fascination that makes snake length important. Size informs bite and fang size, the strength and length with which the snake can strike, and the volume of venom that could be affecting the person bitten.

Notably, all of this information relates less to the snake than it does to whoever meets up with a rattler. Length and weight vary with species and age as well as *raconteur*, but there are what Klauber considers reliable sources (even if present-day sources shake skeptical heads). One such source, W. A. King Jr. of Brownsville, Texas, one of the largest commercial snake dealers in the Southwest, insisted that he measured a western diamond-backed rattler in 1937 as long as seven and a half feet and weighing up to twenty-four pounds with a girth of fifteen inches.[30]

Rattlesnakes have suffered from a poor image for a long time. Ever since Europeans planted their staffs in New World soil, snakes and other carnivores have been on the run. The European adder (*Vipera berus*) found in England, and the horned version described by the Earl of Oxford as Cleopatra's self-inflicted weapon, belong to the same family as their North American counterparts, but the saber-rattling tailed snakes were much more troubling to the continent's new owners.[31] "It is not hard to imagine the consternation of the first settlers in North America when they encountered the pit vipers, in particular the rattlesnakes, which were certainly more abundant then than they are today. The characteristic buzz of the rattle; the sinister mien . . . the forked tongue . . . and the pair of forward-looking 'holes' in front of the eyes, to which it was easy for the superstitious to attribute some malevolent function . . . all combined to make the rattlesnakes, and by association all snakes, the object of fear and dread."[32] The author of this description, a professor of neuroanatomy in Japan, goes on to note that it took science 450 years before the first pioneering work on rattlesnakes was conducted, an astute observation on the disinterest—perhaps, one might even read arrogance—of North America's conquistadors.

Herpetophiles attribute the indifference and disdain for reptile sensibilities to the engrained bias, and even repulsion, against scaled kin: "Humans exhibit less affinity with 'scaly' reptiles compared to 'cute and cuddly' birds and mammals, and many people fear and actively avoid reptiles, especially," notes one, and "few animals have as emotionally charged a relationship with people as do snakes," according to another.[33] The absence of readily comprehended endothermic features in snakes is assumed to mean that reptiles lack associated feelings and cognitive capacities.[34] As Melissa Amarello explains, "Unless you know snakes or other reptiles and have an open mind, it is hard to talk about emotions. Snakes don't have, or at least we aren't able to recognize, the kinds of expressions that dogs and cats have. Snake facial scales can't move much. They don't have movable eyelids. In fact, they have clear eyelids so their eyes are always shut but the covering lid is clear. What a human sees are a blank face and a blank stare—which is very threatening to humans. Snakes also don't have arms and legs or fur—the features that we, as mammals, relate to and use to communicate with each other."[35] We might conjecture that ground squirrels are able to pick up on nuanced facial and body expressions, however.

Such taxonomic chauvinism extends to misperceptions about basic reptile biology. For example, it is widely accepted among American adults and children that snakes eat by unhinging their jaws.[36] In reality, as Harry Greene has demonstrated, snakes accomplish the task by combining two innovative anatomical designs: first, an upper jaw that moves on its own (unlike ours) and a lower jaw, made of a single bone, that opens up much like the elbows of square dancers doing a two-armed "allemande left."[37]

But now an impressive compendium of detailed studies conducted and collected by snake researchers has led to a new image of snakes. Snakes display a variety of behaviors that reflect higher-order faculties such as self-awareness and relational affinities. In addition to superinfrared thermal detection abilities and nuanced tail vocabularies, rattlers "utilize multiple sensory modalities in communication and sociality which include tactile, vibrational, and

Beverly protecting her nest. Photo credit: Roger Repp.

auditory information, chemical information via vomeronasal, olfactory, and taste (taste buds) systems," as well as somatic sense organs.[38] Roger Repp describes an artful method for achieving social deception employed by male rattlers. It is a "behavior called 'stacking,'" which is a form of mate guarding.[39] At aggregate dens of western diamond-backed rattlesnakes, multiple males and females are often confined to a small patch of ground. In an attempt to disguise a coiled female from interloping males, one male will coil over the top of a female so perfectly that only one snake is visible. The female is completely out of sight of the other males—as well as any other onlookers."[40]

Aside from ground squirrels, what other kinds of relationships do rattlesnakes have, and what is known about a rattler's inner world? For many years, even knowledgeable rattlesnake natural historians did not notice these relationships, or were blinded by the prevailing belief that rattlesnakes had no social inclinations aside from mating. Even the existence of a maternal bond was dismissed. "Their propinquity, such as it is, does not result from any maternal solicitude; rather it is only because the refuge sought by the mother is also used as a hiding place by the young."[41] But recent, careful observations have led to the discovery of both rattlesnake social sensibilities and ethical inclinations. Similar to bears and orcas, snakes have social protocol and etiquette that they follow.

Acquiring data on rattler social habits is not easy. Rattlesnakes are notoriously cryptic—they seem a private, shy, or introverted species. It is possible to study rattlesnakes in the more accessible conditions of a lab, and people do, but if the goal is to understand emotions and sociality, it is much more reasonable to do so in a natural setting, where the subject is not threatened and stressed by experimental prodding and poking in a glass-walled tank that in no way resembles natural snake habitat. This was the logic that inspired the method behind Melissa Amarello's serpentine madness, her passion for studying snakes, and rattlesnakes in particular.

In 2001, Amarello moved from Kentucky to the full-blown rattlesnake country of Arizona. There she would put her lifelong fascination with snakes to work in a series of projects on the natural history and conservation of snakes in the southwestern United States and Mexico. After finishing her bachelor's degree in wildlife, watershed, and rangeland resources at the University of Arizona, Melissa continued her studies in a doctoral program in the Biology Department at Arizona State University. There she began looking into rattlesnake social behavior. She and others have discovered that, in addition to interspecies discourse with ground squirrels, snakes have ties that bind.[42] In particular, timber rattlers, one of the largest rattlesnake species, exhibit "key characteristics of other taxa regarded as social [such as] kin recognition, group defense and parental care."[43] Related female snakes who were raised for two and a half years in a laboratory setting gravitated toward and entwined with their litter mates, while those who were not related stayed apart.[44] It is also generally assumed and observed that most snakes are polygamous, but there are verified cases of copperheads (*Agkistrodon contortrix*) bonding in pairs.[45] Other rattlesnakes show similar civil leanings.

Using time-lapse trail cameras that recorded rattlesnake movements and interactions, Amarello and fellow researcher Jeff Smith filmed Arizona black rattlesnakes (*Crotalus cerberus*) engaging in a number of social activities. Black rattlers inhabit the backbone of Arizona's mountainous terrain and get their name from their skin that blackens with time and is crisscrossed with startling orange and white lines.

By quantifying snake-to-snake associations and analyzing the data using social network analysis, a method used to investigate social patterns by characterizing individuals and their relationships to each other via nodes and ties, Amarello and Smith were able to distinguish random encounters from intentional gatherings and who was rubbing tails with whom.[46] The two researchers filmed the Arizona rattlesnakes denning in groups, using communal nest sites, and touching other snakes. Network analysis provided a graphical representation of relationships measured in degrees of one, two, three, and beyond six degrees of separation. The researchers defined the unit of socialization, an association, by proximity. If snakes were resting within one body length from another snake, then an association was tallied. But how, you might ask, did the scientists tell the rattlers apart?

All the usual methods employed by herpetologists to differentiate individual snakes require capture and handling (branding, clipping scales, inserting microchips) both to make the initial mark and to read the mark when the snake is again sighted. But the idea of constant harassment did not align with these scientists' ethics, and it was likely to affect the behavior they hoped to observe. Moreover, since none of these types of markings can be "read" by a camera, filming would be useless. So at first, when individual snakes were found abroad, away from their dens and nests, the researchers caught them and painted their rattles purple and other colors so that they could be readily distinguished and the researchers could collect DNA to measure genetic relatedness. It served as a useful method, but a temporary one because, despite what folk biology says, the number of rattles on a snake's tail does not correspond to age. Rattles come and go. Young rattlers often molt multiple times in a year, and the wear and tear of crawling on the belly and living under rocky shelves tend to break or loosen tail-tip rattles over time.

As they became more familiar with individual snakes, the two biologists changed their tagging methods and modeled them after those used in the study of orcas, elephants, and primates and which do not require repeated handling. The researchers switched to documenting the snakes' unique skin patterns to tell individuals apart and in so doing discovered insights into rattlesnake relationships and emotional bonds. This is how they knew what Priscilla, a beautiful female black rattler, was doing.

Priscilla was a young mother-to-be rattlesnake with whom House, a neonate rattler, was hanging out. House was not Priscilla's son, but Priscilla's care demonstrated that rattlers have sentiments for those who are not genetically related. Although House's mother was elsewhere, the researchers observed Priscilla protectively preventing House from exposing himself dangerously to human predators (the researchers themselves): "Priscilla was pregnant at this time, so we knew that House had a different mother. The two had been resting together under the rock shelter for a while, as was their habit, when House got restless and started crawling around in the shelter. He eventually started crawling toward the open where he would be fully exposed to predators and the hot, summer sun. Priscilla threw her body in front of House, blocking his way out, and turned her head at a sharp right angle (a very unusual posture for a snake) to look directly at him. He immediately turned around and came back to rest in the safety of the shelter, by Priscilla's side. To our knowledge, this was the first observation of 'helping'—where an animal cares for another's offspring—in a snake. Perhaps this is why some female rattlesnakes aggregate during gestation and remain together after giving birth."[47]

Priscilla demonstrated that rattlers not only have family, but friends and "*anti-friends*"—Amarello and Smith's term—that is, individual snakes with whom they choose not to associate. Not all snakes show the same proclivities and personalities. Similar to the variety of white shark personalities, some snakes are gregarious extraverts and others more contemplative introverts:

> We observed consistent patterns but there was also a lot of variation. Within a shared aggregation (hang-out) site, some snakes actively sought each other out, spending up to 57% or as little as 0.5% of their time together, while explicitly avoiding others. Of course, there are environmental factors that influence aggregation or separation—hunting, basking, etc. But our statistical methods accounted for that variation, and patterns of sociality were very pronounced. We saw female-female, female-juvenile,

> and juvenile-juvenile friendships but no male-male friendships (a friendship was defined as a pair of snakes that spent more than twice as much time together than if they were associating randomly). The males didn't form strong bonds but this did not mean they were hostile because we often observed them associating together; they just did not get that close.[48]

Amarello's point about intraspecific differences echoes those of Repp and others who bring attention to how individuality can be noticed at many levels of social organization. There are not only huge differences among rattlesnake species, but also variations in habits, culture, and personality from population to population. Similar to the linguistic differences between Norwegians and inhabitants of the Italian Alps, rattlesnake "ethological dialects" differ in response to variations in microclimatic and micro-terrain.[49] Evocative of orca sympatric conservatism, rattlesnakes appear to prefer to keep company with their own. Roger Repp describes:

> To illustrate how complex and distinct rattlers are, I'll give you an example with the western diamond-backed, who we just refer to by Latin species name, *atrox.* In contrast to *cerberus,* who birth their young within ten meters of where they den and where rattlesnakes will "help out" with their sister's babies, *atrox* do not form rookeries. They like to be on their own, and there is seldom a male around when there are babies. Furthermore, once babies have shed, they go off on their own, and as far as we know they never get together with their mother or family again. It is possible that they do, but not at the den sites that we have witnessed. But, this in no way relates to a lack of emotions and bonds. I saw a male diamondback travel 100 meters, three times in the dead of winter, to be with and attend to a sick female. We have had as many as ten male *atrox* under watch throughout thirteen winters, and never have any left a den and returned in winter.[50]

Repp speculates that the sick female was a former mate who normally overwintered with the male.

Male Rattlesnake in embrace with dead female. Photo credit: Roger Repp.

Repp also describes a photograph taken by herpetologist Gordon Schuett showing a male sidewinder embracing a dead female who had been killed by a car. He is hesitant to call it grief, yet an embrace that is sometimes mixed with sexual overtures is not uncommon among other species such as elephants who "incredulous of despair, Half-taught in anguish, through the midnight air, Beat upward to God's throne in loud access, Of shrieking and reproach," desperately try to resuscitate by lifting and sometimes mounting a dying friend.[51]

These observations, as similarly noted by Fred Buyle with sharks and Charlie Russell with bears, underscore the significance of individual differences in personalities and values and their dynamic natures. Neuropsychologist Darcia Narvaez describes a "multiple-ethics theory [that] can help explain the variability in moral functioning that we see in ourselves." At the species level, neurobiological development of moral motivation "emerges from early experience to shape long-term well-being and moral orientation" and so we can expect to observe general patterns of behavior and mental and emotional states. "But," and here neuropsychology generally departs from conventional wildlife biology, "on a moment to moment basis, an individual's morality is a shifting landscape. We move in and out of different ethics based on the social context, our mood, filters, stress response, ideals, goals of the moment, and so on."[52] This perspective cautions us not to overlook individuality in the search for generalizations. How an interaction at one point in time unfolds is a function not only of general species' behavior, but also of conditions internal and external to the subjects. As we shall see, this multidimensional view is in step with the teachings of ethology's forefathers, Konrad Lorenz and Niko Tinbergen, who were superb observers of nature.

Roger Repp recalls another situation that illustrates snake emotions and strong feelings of attachment. Using radiotelemetry, Repp and a colleague were able to chart detailed movements of a female *atrox,* affectionately named Ali, for two years. By late spring 2006, it was clear that she was pregnant. Distal portions of her body bore the telltale plumpness of a mother-to-be, and by early August, Ali had settled into a series of rat-scat-infested holes beneath a well-armed prickly pear cactus.

For the next three weeks, Ali remained unseen. If it had not been for the transmitters, Repp would have been unable to confirm that she was hidden within. Then, one night, while tip-toeing around to find out the progress of Ali's pregnancy, he came upon her on open ground more than fifty meters west of the prickly pear nest site. Given her lean look, Roger deduced that she had given birth, and given her distance from the nest site, he assumed that the snakelettes had shed their skins and were on their own. Shed skins would be handy for DNA fingerprinting so Roger proceeded onward to the nest site.

The trek over to the nest wasn't easy. It was a dark, overcast night with navigation aided only by a compass and a temperamental GPS unit. As Roger walked in the direction of the nest site, Ali pulled an "end around "and placed herself directly in his path. A second end around put Ali smack between two palo verde trees (*Cercidium floridum*). This coiled, determined snake book-ended by unfriendly spiny trees made a formidable barrier to Roger's passage. The only possible way through was via the spot where Ali had herself lodged.

Repp tried to gently push her out of the way with his snake hook. But no go. The mother snake refused to budge. This was totally out of character for *atrox* and other rattlesnakes. "I was amazed," recounts Repp. "Eleven times out of ten, they will either fight or flee when provoked and Ali did neither, she was doing her all out best to block me and make sure I did not get past. She hugged the ground so tightly that she turned into what I can only describe as a snake frisbee!"[53] He carefully picked Ali up with his snake tong, and placed her behind him, her "frisbee" posture remaining intact, then continued on to the prickly pear nest site. But when Repp arrived on site, there was no sign of any baby snakes or their skins. He skirted the prickly pear fortress and peered deep down into each of the holes. Nothing and no one. He stood up and was making some notes on the data-sheet when he spied her. There she was, Mama snake, frisbeed again and all. Ali had slither-stalked him all the way back to the nest site, a distance of fifty meters or so.

Thirty-six hours later, when Repp and his colleague, evolutionary biologist and herpetologist Gordon Schuett, visited the nest site, they saw the traces of Ali's worry: four shed skins of neonatal *atrox.* In the researcher's mind, Ali's lack of flight was firm evidence of motherly love and concern. She had obviously been out hunting, leaving her youngsters safely tucked in the recesses of the nest, when the intruding human came upon her. When asked why she had not chosen the usual rattlesnake alternative to flight, that is, rattle, lunge, and bite, and whether her restraint toward the man indicated affection, Repp paused, then answered: "Perhaps through the years Ali has come to learn the futility of fighting me. She instead chose a more passive approach to resistance by trying to block my path. I won't say Ali loves me. I'm not even sure she likes me. But I do think that she knows who I am."

Now let us pause a moment and view this scenario from the vantage point of psychology. According to Daniel Goleman's definitions of emotional intelligence, Ali, similar to the Morelet's crocodile, fulfilled all four of the elements. She showed an ability to manage strong emotions that, as evinced by her behavior, arose out of concern for her young (she regulated her affect), she engaged in active measured evaluation and decision-making in terms of who the human was and what he was doing (showing self-awareness and empathy), and she acted appropriately in a specific situation based on this

Rattlesnake combat dance. Photo credit: Roger Repp.

judgment (she integrated all of this information appropriately). Over the period of their study, the rattlesnake had learned who Repp was and what he was up to, or rather, what he was *not* up to. She used these data to inform and modify her risk perception and response. In short, Ali scores high in emotional intelligence. The fact that Ali learned, then weighed her learning against the odds, shows intelligence and sensitivity; it demonstrates that she's someone who is willing to step beyond inherited, and well-justified, prejudice against humans.

Female rattlesnakes are not the only ones with complex social interactions. Male snake society and emotional terrain are also nuanced. For many years, it was believed that two rattlesnakes standing erect and swaying in a somatic duet were engaging in a mating ritual. But in fact the ritual involves two males. Klauber describes it is as "an affair wherein the obvious isn't the truth, and the truth is stranger than the obvious . . . But the stylistic gyration of the performers suggests a symbolic dance, and as the term 'dance' has been applied to the exhibition by those who have most often observed it, hence, it shall be; hereafter without the quotation marks."[54]

Although Klauber maintained that the combat dance, as it is known, always ended in a draw, more recent observations show that there is usually a definitive winner and loser. Notably, however, both snakes use an amazing level of restraint: no damaging or lethal violence is exercised. After being driven from the den by the winner, the losing male may return within a few hours or days to try again. "The beauty of this is that rattlesnakes make war but do not hurt their opponent."[55] Yet another expression that is reminiscent of the pacific orca.

As Repp asserts, there is a whole world of subtle and intricate inter-snake exchanges that can occur. A film clip of a very large *atrox* named Tyson at one of the dens provides an example:

> Tyson was omnipresent . . . and always on the move. He was alert and aggressively took note of every nook, cranny, and snake on his turf. He appeared to have complete command of his world. To be sure, the film crew caught mating and fighting, but that is not what got me out of my chair. What truly piqued my attention was the intense tail waving that transpired. In one sequence, Tyson was investigating (rapid tongue flicking) a cluster of about 15 adult conspecifics. As he crawls on top of the pile, four tails immediately rose out and began waving back-and-forth in sinusoidal fashion, not unlike caudal luring. Similar behavior occurs in male copperheads during dominant-subordinate episodes or bouts. Tyson zeroed in on one of those tails and used his snout to push and extract the snake that owned it. It turns out this other snake was a male, but not quite as big as Tyson. The excellent footage clearly showed that his rattles were tapered and complete—a younger snake, perhaps an "upstart." Once Upstart caught wind that he was noticed by Tyson, he rapidly fled the pile, heading downslope and jetting toward a sandy wash positioned south of the cement-like rocks. Tyson overtook him,

Tyson, Western Diamondback Rattlesnake. Photo credit: Roger Repp.

> sprawled over top of him, and pinned him on the sandy sand stratum. When Upstart began waving his tail back and forth, Tyson reacted by grinding Upstart deeper into the soft sand beneath him. Male dominance was clearly exhibited without sparring in outright combat with neck-to-neck vertical postures. Now, I'm out of my chair, exclaiming, "Whoa, can we rewind that?" One of the three film crew members asked me, "Is that tail-waving a sign of submission?" I had to admit that I had never seen that behavior before, but submission certainly seemed like a likely explanation.[56]

Repp describes another complex male-male exchange involving the mighty Tyson.

> He was coiled on top of a group of several other *atrox* at the eastern-most entrance to the den. Only the front two-thirds of

> his body were visible. The rear third of his body was buried and groping into the cluster beneath him. While he was doing this, a young, smaller adult male *atrox* emerged from the depths of the den and coiled beside him. Tyson showed great interest in this new development and immediately began to chin rub on the coiled form. In almost nonchalant fashion, the new arrival pulled his tail from beneath his coils, and methodically waved it back and forth. His tail was positioned in such a way that it was directly in front of Tyson's face. Tyson's reaction was swift and fierce. He ascended half his body length above the new arrival and came crashing back down on him. As Tyson crashed on that snake the thud was audible—it was a firm smacking. Without further encouragement, the new arrival shot back down into the den, out of sight in one flat second. Tyson gave a limited chase for half a body length, and returned to the cluster of snakes beneath him as if nothing had happened . . . What had transpired between these male snakes? Something like: "No, idiot, you got me all wrong. You can't mate with me—I'm a male!" Rather than respond with his tail, Tyson used "dominant" body language to let his intentions be known. "Oh, really . . . you're another male? Then you better get out of here!" The new male got the message, and off he went. However simple or complex it was, these two snakes communicated with each other. A message was delivered and received—loud and clear.[57]

Oh to be Harry Potter, a parseltongue, one who is able to understand and speak with snakes!

When rattlesnake studies are passed through the relational lens of attachment theory, an intriguing picture of these species materializes and begs some tantalizing questions. Rattlers share traits associated with mammalian species known to be social: they are long-lived (timber rattlesnakes live up to thirty years), late sexual bloomers (timber rattlesnakes take up to nine years to mature sexually), tend to cluster at the rattlesnake-equivalent of watering holes, and show mothering personalities. Like elephants, rattlers are viviparous, meaning that little snakes are born live. Baby rattlers develop inside their mother, from whom they emerge, living, birthed to the outside world in litters of one to more than a couple dozen. Gestation takes between three to five months depending on the species. On average, rattlesnakes spend about 8 to 10 percent of their lives in a developmental context that appears to be designed for fostering a secure attachment. This envelope of protection and nurturance is comprised of a mother in a communal habitat that may have been shared generation after generation, give or take the flux of immigration. Beyond this first stage, young snakes stay near home for a couple of weeks until they undergo a first molt.

Sociality studies suggest that young rattlers graduate to another developmental stage: they stay within the broader complex used for denning or communal hunting, which is populated by adult rattlesnakes, while still under the oversight of diligent mothers or helper females such as Priscilla. In theory, a tiered socioecological structure provides a way for the reptiles to tune their version of the mammalian right hemisphere's socioaffective centers—the areas key to stress regulation, self-development, and affiliative behavior. In other words, baby rattlers are cultivated in rattlesnake morals and ethics. Darwin would probably agree. As he so famously wrote: "The following proposition seems to me in a high degree probable—namely, that any animal whatever, endowed with well-marked social instincts, the parental and filial affections being here included, would inevitably acquire a moral sense or conscience, as soon as its intellectual powers had become as well, or nearly as well developed, as in man. For, firstly, the social instincts lead an animal to take pleasure in the society of its fellows, to feel a certain amount of sympathy with them, and to perform various services for them."[58] Bears learn bear manners, crocodiles learn crocodile manners, and so with rattlers. They also learn neighborhood rules: in the case of rattlesnakes these rules include those of ground squirrels.

In addition to emotional intelligence, Priscilla and Ali displayed traits and skills associated with what comparative psychologist and educator Edward Thorndike called *social intelligence.* Thorndike, perhaps best known for his development of the "law of effects" (in brief, the benefits of positive versus negative reinforcement in learning), identified three main areas of mental development: abstract intelligence (the ability to grasp concepts and ideas), mechanical intelligence (the ability to manipulate objects in the biophysical environment), and social intelligence (the ability to navigate interactions with others). Since then, evolutionary psychologists and neuroscientists have expanded beyond what was once a purely biological view to consider the role of social knowledge and skills in survival and day-to-day life. From the vantage of comparative neuroanatomy, it should come as no surprise that the "brains of social organisms have neurobiological circuits that recognize, compute, and manipulate socially relevant information." Indeed, further research such as polyvagal theory developed by Stephen Porges, Director of the Brain-Body Center, University of Illinois, Chicago, illustrates the intricate relationships between neural substrates, stress regulation, and sociality.[59]

For many years, attention focused on primates whose brains are, relative to body size, very large compared to other vertebrates. This observation, a correlation between large brain size and social-system complexity and group size, led to the "social brain hypothesis" put forward by anthropologist Robin Dunbar. The idea is that humans and other primate brains evolved to their present sizes in order to solve not just complicated ecological problems, but also social ones.[60] This concept has been extended to include other mammals and birds, but the relationship in these cases is somewhat modified. In the case of primates, "there is a quantitative relationship between brain size and social group size (group size is a monotonic function of brain size), presumably because the cognitive demands of sociality place a constraint on the number of individuals that can be maintained in a coherent group." By contrast, for other mammals and birds, the relationships between brain size and social group size are qualitative: in particular, "large brains are associated with categorical differences in mating systems, with species that have pair-bonded mating systems having the largest brains" (pair-bonding avians include such species as ravens and Amazon parrots).[61]

When asked about the relevance of the social brain hypothesis, which asserts that bigger brains evolved to address complex sociality, Rockefeller University's Erich Jarvis responded from the perspective of current neuroscience: "I can imagine how more complex social interactions select for more complex behavioral traits. But this does not mean that they select for bigger brains. Could this extend to reptiles? I don't see why not."[62] Which is exactly what Amarello and Smith's discovery of rattlesnake sociality suggests. It looks like once more, neuroscience, in combination with astute field observations, may lead another taxonomic branch, in this case, reptiles, to soon receive the regard that other species with socially minded brains enjoy.

Now here come the questions: Is it during this second stage of socialization that young snakes get to look at and learn about hunting techniques and subtleties such as ground squirrel tail communiqués? Can some of the differences in personality that Amarello and Smith observed be ascribed to different developmental contexts and attachment styles? What are the natures of rattlesnake culture and ethics? And do elder rattlesnakes assume the roles of matriarch and patriarch as observed in species such as elephants and orcas? These are just a few queries that pop up when neuropsychology is applied to rattlesnakes. As yet, there are no direct answers, but some will certainly be forthcoming given the new paradigm that herpetologists have adopted.

But as someone who studies the socioaffective lives of rattlesnakes, Melissa Amarello can attest to what neuroscience predicts: snakes are extremely affected by stress and feel fear and the specter of death as profoundly as our own species does. Using biological sense, she rectifies the image of a rattler from an evil menace just waiting to bite you to a vulnerable, shy, and scared armless, legless soul who is just trying to get along:

> You can tell a lot about how a rattler is feeling by just watching. Fear is one of the easiest emotions for us to pick up in others. But people have been told that what is really a sign of fear in a snake is aggression. They just assume that when a rattler is rattling, it means that the snake is being evil and about to strike. But rattling is not a threat. From the snake's perspective, once he starts to rattle, he thinks that he has been seen and that he is about to be killed. He is not preparing to attack. You never see snakes rattle when they are going to kill. It makes sense. Why would you give yourself away if you wanted to hurt or harm someone? Rattlesnakes are terrified of humans. If you start to mess with one, they often end up just hiding their head under their coils. I have only seen them do this with humans—like at the round-up. They hide their heads and just give up. It is really sad to see an animal give up on life.[63]

The round-ups to which Amarello refers are annual gatherings where thousands of people participate in the killing of thousands of rattlesnakes. The Sweetwater, Texas, round-up is the largest of several held at other locales in the United States. Weekend events can draw up to 35,000 spectators to watch more than 16,000 rattlesnakes die. At the 2016 Sweetwater round-up, 21,000 snakes died.[64] By the time the snakes reach a round-up, they have been flushed from their dens with gasoline, kept in cages that are too cold or too hot, and not given food or water for weeks and sometimes months. Furthermore, they have been piled into cages so crowded that many asphyxiate.[65] Those who survive are used in various shows and activities. The Skinning Pit demonstrates to families, including children, how to behead and skin a snake, and gives them the opportunity to skin their own snake. In other events, audience members are given machetes and taught how to skin the just-slain snake themselves. There is also a Miss Snake Charmer pageant, where contestants must kill and skin a rattlesnake.[66] At another popular booth, participants can have the unique experience of having their photo taken with a live rattlesnake draped around their shoulders. The experience may be titillating, but there is no danger of being bitten either, because it is reported that the snake's fangs have either been torn out and her mouth sewn shut, or some other method has been used to prevent a bite. Elsewhere, snakes are killed on the spot and served up for lunch, with diners inevitably commenting, "It tastes just like chicken!"[67]

The snakes await their fates in cages where they experience various stages of mental and emotional collapse. Melissa describes the typical state that a snake is in when they first arrive: "There was one individual snake who touched me. He was in the first stage at a round-up where they take snakes out and measure them. He had already been handled roughly, his face was bruised, swollen and was bleeding and the worst was yet to come." The air is literally abuzz. To most modern ears, the hollow rattle of a snake about to be beheaded conveys little information. But not for the social snake researcher: "When I hear a thousand rattlesnakes rattling, I hear a thousand snakes screaming."[68] This is entirely consistent with what neuropsychology predicts when someone is rounded up, brutally assaulted, and taken from their home

Thousands of Rattlesnakes are killed annually for entertainment at various "round-ups."
Photo credit: Jo-Anne McArthur.

and family with the growing realization that no one will come to save them, and nothing but death awaits.

Harry Greene shares a gentler experience of rattlesnake psychological vulnerability. Having studied snakes for decades, Greene and his colleagues developed a keen awareness of snake sensibilities. As a result, they have switched from macho techniques that pin and catch rattlers to softer ways mindful of the stress a captured snake endures:

> We were at pains to minimize our impact on the snakes, partly because of personal values with regard to how animals should be treated, partly because in terms of watching normal snake behavior, traumatizing them would have obviously been counterproductive. Thus, we gently lifted snakes into a bucket for capture and never used any sort of pressure-restraint (manual or otherwise), and restrained them for anesthesia by gently coaxing them into a plastic tube. I think it is likely they experienced some sort of post-surgical discomfort from the incision for implanting a transmitter, although I didn't observe any overt signs of that. We considered use of some sort of analgesic but decided against that because: a) at the time at least there was no evidence that available pain-killer would have any effect on a snake, and b) tenderness is adaptive in the sense of causing one (us, snake, whatever) to favor a wound and not otherwise bump the snake, etc. We wanted the snakes to take it easy so as to promote healing. I have no doubt that rattlers are very familiar with many specific attributes of their home ranges so we reasoned that being back in familiar surroundings would be less stressful for the snake than in a cage. Subsequently, our protocol was to release them at the exact site of capture within ~24 hrs. of surgery. Typical behavior for a snake upon release was to crawl a short distance to some secure site, remain inactive for a few days, and then resume seemingly normal behavior.[69]

Greene believes that this respectful way of interacting, even if the snake was only tolerating their overtures, shifts perceptions from aggression and fear to cooperation. He describes one incident that he thinks reflects how consistent, respectful approaches to snakes significantly alter the snake's own perception of danger:

> With David Hardy, my anesthesiologist collaborator, I watched one black-tailed rattlesnake for twelve years, encompassing 569 encounters, five telemetry transmitter implantation surgeries, and four pregnancies with subsequent maternal care. Our handling and observation protocols emphasized not traumatizing the snakes, and I believe I never heard that animal ("Super Female 21") rattle in all of our encounters, including capture for surgery (which was done under anesthesia). Interestingly, to say the least, I once inadvertently ended up with my face within inches of her head (she was uncharacteristically, I thought at the time, up in a tree, under which I was sitting trying to figure out where she was), and she didn't so much as flick her tongue, let alone rattle or strike. I do smile to think of that as a sort of Androcles and the Lion moment, as I very much like rattlesnakes—and it is tempting to wonder if I might not have had a very bad bite, *if* she'd been taught by rough handling.[70]

This account stands in astounding contrast to the conventional snake-as-aggressor lens. When I suggested to Greene that his account showed that tensions and trauma associated with capture, surgery, and other human manipulation were mitigated by the relationship that had developed between snake and man, even to the point that Super Female 21 had developed a profound trust in her handlers and bonds, Greene responded with some reserve, "I'm cool with referring to me and Super Female 21 having a 'relationship,' but it feels like quite a stretch to refer to her as having 'profound trust' in me, or to us having a 'bond.' One can use those words to label what I described with the rattlesnake, and I was undeniably fond of her, but my caution is occasioned by strong doubts that the 'bond' between me and the favorite dog of my life, let alone with a lover or best friend, is in any meaningful sense the same as my relationship with that rattlesnake."[71] These are early days in social snake studies, however, and with neuroscience's ever-

Rattlesnake skinning contest at round-up. Photo credit: Jo-Anne McArthur.

expanding taxonomic net, such human-snake interactions may be up for re-interpretation.

Amarello provides another example of cross-species prosocial rattlesnake personalities. Henry was (and Melissa adds, "I hope *is*") a large, old western diamond-backed rattlesnake. She met him doing fieldwork over a couple of years:

> Other researchers have said the same things, that the older and larger snakes are the most mellow, sweeter, they don't rattle much, and they are very tolerant of people. Henry was very tolerant of people because he was well treated by people. He seemed to trust us and he would let us get close. I don't remember ever hearing him rattle even though we lived right around him. On one occasion, I was taking pictures when Henry started to crawl closer and closer to Jeff who was nearby sitting on a tree branch. If Jeff [Smith] hadn't gotten up, it seemed like Henry would have crawled right onto his lap! You have to appreciate what this means

> to a rattler. When they are out moving around they are most exposed to predators, so usually at their most nervous, but he apparently felt relaxed enough to risk approaching us. Another time, Henry was fighting with another male, doing their combat dance and they got so close to me that my camera wouldn't focus so I had to keep backing up to prevent them from running into me. He even ate in front of us once when we gave him a dead dove who had died flying into a window. That's another vulnerable moment for a snake, if their mouth is full, they can't bite. Yes, I'd say we had a friendship, and a meaningful one. He seemed to recognize us both, approached us at times in a curious, rather than defensive manner, and never showed much of suspicion or fear.[72]

If these exchanges had taken place between two mammals, that "something" between snake and human would evoke descriptions such as intelligence and love. Whatever the interpretation, something implicit in the snake-human relationship in the cases of Greene, Repp, Smith, and Amarello overrode what is understood to be the rattlesnake species' common reaction to danger and fear, namely the "eleven times out of ten" response of flight or fight.

So what we find is that snakes have meaningful connections to their physical and social environments—that is, they demonstrate what would be called, in the case of tribal and indigenous humans such as the Quechua, "ecocentric attachment." In contrast to Western or urbanized cultures, tribal peoples do not usually draw the line between plants, the stars, rivers, and other "biophysical" entities, let alone between species. They, like snakes, are bonded to those who sustain them and this includes, in fact is defined by, the land, waters, skies, plants, and animals with whom they live. There is growing evidence that rattlesnakes, like elephants, are vulnerable to displacement from their homes and social group. This has become an issue of late with an increase in wildlife translocation.

There are two main purposes for moving a snake—reintroduction and mitigation. Alarmed by plummeting wildlife populations, conservationists are trying to stem the hemorrhaging by developing captive breeding pro-

Melissa Amarello and Henry. Photo credit: Jeff Smith and Melissa Amarello.

grams. The idea is to use and reintroduce captive-bred endangered species to prevent extinction. Many efforts, however, fail. The reasons for failure are multiple, complex, political, and biological. As we saw earlier in the discussion of grizzlies, captive breeding efforts undertaken for the Californian condor (*Gymnogyps californianus*) often did not take into account normative attachment and developmental needs, with the result that young animals were ill prepared for living on their own when they were reintroduced into the wild. In addition, the human practices that have brought a species close to extinction have not changed, which means that the environment where the captive is reintroduced may well have inadequate social and ecological resources for survival.[73] Reintroduced individuals have no special immunity from hunting; urban, suburban, and farming development; power lines; pesticides; and other threats.

The second kind of transport is mitigation translocation—that is, ridding human communities of "nuisance" wildlife by moving individuals or groups. Escalating human population and concomitant habitat appropriation have led to increased human-wildlife encounters where the inevitable outcome is that the animals are killed or there are demands for their relocation. Urbanized folk may enjoy nature shows, but they are often loathe to share their backyards with skunks, raccoons, bears, rattlers, and other neighbors. The so-called trespassers are usually summarily trapped, shot, or poisoned. Although mitigation translocation is seen as a more benign alternative and a way to support conservation of the species, it is nonetheless problematic.

Erica Nowak, a Northern Arizona University herpetologist, reports that 57 percent of western diamondbacks died in a translocation effort. She considers that this holds true in other cases. Rattlesnake mortality is particularly high if relocation occurs outside the home range (which she suggests is, on average, one square kilometer) because long-lived reptiles such as desert tortoises (*Gopherus agassizii* and *Gopherus morafkai*), Gila monsters (*Heloderma suspectum*), and rattlesnakes "are intimately connected both to the landscape and to one another."[74] Rattlesnakes will try and make it back home to their natal and social circles. This exposes them to predators, highways, unfamiliar environments, and other stressors. Imagine yourself trying to get home with no wallet, no credit card, no rental car, no cell phone or tablet, and no food, all while having to navigate through a maze of hostile neighborhoods—and you have just lost your prescription glasses.

Recall that the infrared studies showed that pit vipers can "distinguish cool moving objects against a warmer background and warm moving objects against a cooler background" via a thermophysiological mechanism that enables these species to find and eat endo- and ectothermic foods of choice such as frogs. To accurately detect the delicate thermal signatures of their prey, rattlesnakes need to be able to discern the signal (viable prey) from the noise (environmental background), which implies some sort of finely tuned calibration. Rattlesnakes are living, breathing thermal filters exquisitely calibrated to their habitats. So when their habitat disappears and they are tossed somewhere completely unfamiliar, possibly with other snakes who are already strapped for food, it is likely that they experience tremendous physical and emotional disorientation. The snakes have been torn from land, family, and friends, and given a wonky compass to boot. In the language of neuropsychology, rattlesnake translocation and relocation are traumatic.

Rattlesnake emotional and social intelligence raises some interesting ethical, pedagogical—and existential—questions. The union of neuroscience and field observations presents us with a flavor, conceptually, of what Val Plumwood experienced—an abrupt demotion of our species from its Olympian heights to the earthy serpentine realm. Suddenly, we find "ourselves in mutual, ecological terms, as part of the food chain, eaten as well as eater."[75] Since modern human identity has been largely defined by what and who we are not—in this case, snakes—then when snakes turn out not who they are supposed to be, neither are we.[76] The resulting identity confusion may be somewhat akin to the disorientation that a trapped and relocated rattlesnake feels.

Given the new view on reptiles, Schuett and his colleagues are calling for a new paradigm of snake sociality.[77] It is essential, they assert, to break "conceptual barriers that concern the complexity and importance of social behavior in snakes." Standard scientific terminology cannot capture all of these observed subtleties. For one, subjective emotions and mental states of nonhuman animals, and in particular, non-mammals, while firmly established in neuroscience, have yet to be legitimized in ethology, wildlife biology, and related fields. Thus, there is a reluctance to see, let alone, describe, snake affect and social proclivities. Neither is science, as commonly articulated today, equipped to capture what it has conventionally defined as nonexistent—the tender glow of motherhood in the face of a white shark, cutting genius of a crocodile at work, concerned look of a helper rattler protecting her friend's child, and careful assessment by a grizzly when confronted by intruding humans. The graceful mystery of male rattlesnakes' combat dance, the protective uncoiling of a pregnant mother snake, and the death rattle of snakes suffocating in tiny cages cannot be contained in the neat container of an ethogram (the inventory that is created to catalogue a species' sets of behaviors). The careful narratives of Amarello, Smith, Greene, Rundus, and Repp illustrate that it takes an open mind and keen eye as well as an expanded, more relaxed, conceptual and linguistic vocabulary to faithfully communicate the nuances of the complex visual narrative embodied in relational observation. As James Blachowicz, professor of philosophy at Loyola University Chicago, notes, scientific investigation and the creation of poetry have much in common.[78]

Subsequently, just as we are more than the sum of our individual actions, so are rattlesnakes, alligators, Komodo lizards, and their kin—and behavior is only one card in psyche's deck of expressions whose interpretation is relational and context-dependent. To understand what a snake is doing and why, biologists must learn the skills and sensibilities of psychologists and empaths.

Like any other merger, science's transition to a new, social snake paradigm may start off a bit awkwardly.[79] By definition, paradigm shifts are more than a change of theory. They demand a change in how the world is perceived and, as Plumwood could attest, a change in how we ourselves are viewed. But the joining of animal behavior and human psychology has perhaps something on its side that makes the convergence easier—the two fields share a common bloodline.

In large part, animal behavior studies grew from the ways and work of natural historians who comfortably traveled in a very free intellectual space. Progenitors of modern ethology reveled in exploring qualities and attributes that spanned humans and nonhumans alike. Prior to the second European war, lines between academic fields were much more fluid. Psychologist John Bowlby and ethologist Nikolaas Tinbergen eagerly traded discoveries with scant notice of species' boundaries, in much the same way that quantum physicist Wolfgang Pauli and psychologist C. G. Jung exchanged ideas about the mind and universe. In part, this was because students of the natural world in the days of Konrad Lorenz generally lived much closer to nature than their successors would. Much of nineteenth and the early twentieth century inquiry, before knowledge was fragmented into finer and finer areas of specialization, was conducted with what Tinbergen called, in his 1973

Nobel Prize acceptance speech, *scientia amabilis,* a "loving science" and scholarship.

The Dutch scientist's observational skills were not cultivated in the classroom or from the pages of a book but during childhood wanderings through the woods and waters of his native countryside. For many years, Tinbergen eschewed the formal path. It is hard to believe that a future Nobel Prize winner would be reluctant to enter formal academic study, but it's true. Eventually, at the urging of his father, who was a lettered scholar himself, Tinbergen completed a doctorate at Leiden University and went on to enjoy a long sojourn of research that spawned many prominent ethologists. It is telling that his 1932 dissertation, in biology no less, is the shortest on record at his alma mater.

Not all, however, was love and light. There was also a good deal of capturing and killing, in the name of science, the animals whom nature researchers purported to love. Gerald Durrell, brother of well-known writer Lawrence Durrell, and himself an author of multiple autobiographical books on escapades with wildlife, made a living by catching, transporting, and selling wildlife around the world to zoos and various collections. Today much of modern science follows the same practice and often harms those it seeks to save. Although current rattlesnake endangerment comes from efforts to rid areas of "nuisance" wildlife, widespread habitat destruction, private collectors, and round-ups, it seems that "prior to the 1960's, researchers may have constituted the greatest human threat to twin-spotted rattlesnake populations."[80]

Tinbergen witnessed, with some sadness, the change in science's methods and means. When World War II ended, and he returned from a two-year stint in a German prison, science and society were very different. In an era of heady hubris inspired by the atom's conquest, the momentum and mechanization of the wars were now turned toward science. Animals were natural targets for the mission and were increasingly drawn into the vortex of research bent on progress. Cows, pigs, sheep, and chickens were pulled from pastoral portraitures and inserted into the business of modern, mechanized farming. Cats, dogs, rabbits, chimpanzees, monkeys, and other species of wildlife were recruited press-gang style to fatten research's growing midriff.

Medicine redoubled its search for better and cheaper "models," animals upon whom, as human surrogates, scientists could perform all sorts of procedures.

Nonhuman animals became a staple in the growing field of psychology and its experiments and studies. In 1907, psychologist Robert Yerkes published a book, *The Dancing Mouse,* to promote the virtues of the rodent as a model laboratory subject, but victims of science's pursuits were not limited to *Rodentia.* The primate center, now in its eighth decade, that bears Yerkes's name, boasts a collection of 3,400 primates and 12,000 rodents who are "critical to the Center's research in the fields of microbiology and immunology, neurologic diseases, neuropharmacology, behavioral, cognitive and developmental neuroscience, and psychiatric disorders . . . [and ways] to: develop vaccines for infectious and noninfectious diseases; treat drug addiction; interpret brain activity through imaging; increase understanding of progressive illnesses such as Alzheimer's and Parkinson's diseases; unlock the secrets of memory; determine how the interaction between genetics and society shape who we are; and advance knowledge about the evolutionary links between biology and behavior."[81] Animals became, and continue to be, viewed as an indispensable, sacrificial means in the search to understand ourselves.

The demand for control-based scholarship clashed with the more hands-off approach made famous by such celebrated natural historians as John Muir. Already parsed into classification schema, animal lives were further reduced by statistics and behavioral inventories. The graceful lilt of Thompson's gazelle and bottlenose dolphin's undulation were charted, fitted, and boxed into comestible units for collective consumption. Species after species was transformed into anonymous data referenced only with a number and an antiseptic Latin moniker. Jane Goodall's naming of her chimpanzee subjects recalled earlier times of Konrad Lorenz and his capuchin monkey named Gloria, when a more intimate practice was not uncommon. Life changed for the nature aficionado, too.

Spending time in nature became increasingly less affordable and was even considered unnecessary. Ethology, zoology, and other animal related -ologies provided one of the few ways to legitimately and viably mingle with the great outdoors. Gone were the desultory days of the genteel scientist. There was no time to muse with residents of the woods and waters. Writer Michael Lind compares the lyrical work of naturalist Loren Eiseley with present-day scholarship: "Before the rise of a self-conscious intelligentsia, most educated people—as well as the unlettered majority—spent most of their time in the countryside or, if they lived in cities, were a few blocks away from farmland or wilderness . . . I suspect that thinkers who live in sealed, air-conditioned boxes and work by artificial light (I am one) are as unnatural as apes in cages at zoos."[82] In the postwar West, governments vigorously began turning a funding crank that churned out a stream of "brave new world scientists" quite different from scholars of yesteryear. Successive generations formed an impenetrable phalanx trained to do what Francis Bacon had purportedly suggested: put nature on the rack "to be studied through interrogation."[83]

Such muscular research was unsettling and foreign to Tinbergen's naturalistic approach. When he met with crossover psychologists cum ethologists such as Yerkes, Tinbergen confessed, "I was frankly bewildered by what I saw of American Psychology."[84] He pressed for bringing back the "old" method of "watching and wondering."[85] In his words, "I call the method old because it must have already been highly developed in our ancestral hunter-gatherers, and it still is in non-westernized hunter-gatherer tribes such as the Bushmen, the Eskimo, and the Australian Aborigines. As a scientific method applied to Man, it could be said that it was revived first by Charles Darwin in the 1872, *The Expression of the Emotions in Man and the Animals,* London, John Murray."[86] As the twentieth century progressed, some connection to field-based studies was retained, but active manipulation, with its angular demands, dominated.

This creates a conundrum for many. "Life" sciences too often involve the death and killing of the animals who are the subject and object of the original passion. Melissa Amarello recalls her encounter with this imperative:

> For my Ph.D. project, I was asked to come up with a manipulative experiment. Not because there was a particular experiment that the committee thought would support or refute a potential hypothesis about rattlesnake social behavior, but just because a manipulative experiment was considered an essential ingredient

> in doctoral-level research whether it was needed or not. One of the things they proposed I do was to relocate some of the snakes from one social group to another group. This probably doesn't sound like a big deal, but relocating adult snakes decreases their chances of survival by about 50%. So not only did this seem unethical to me, it also seemed like a stupid experiment if, regardless of whatever happened related to sociality, half of the animals would die. If I had chosen to do an experiment in the lab, all of the snakes who participated would have ended up in the freezer. It was chock full of stacks of coiled, dead snakes, no longer needed after the study was completed.[87]

Her recollections recall Tinbergen's "bewilderment" when he encountered American science: "At some point I realized that my love for the animals whom I studied made me different—I was the strange one who wouldn't consider a lethal experiment. My whole point of going to graduate school was to help snakes. I thought I could make a difference for them. No one seemed particularly interested in how their research would help the animals involved. When I started out, I truly believed that that was what science was for: helping wildlife. But in many cases the animals are just tools used to answer a question. Even though I only chose experiments that weren't lethal themselves, normally those animals would be killed at the end anyway."[88]

Since by mandate of the Arizona Game and Fish Department snakes who are captive-born or live-caught cannot be released into the wild at study's end, Amarello took the former research snakes home: "Government regulations don't allow animals to be released back into the wild after they've been in captivity for more than a few days. The fear is that the snakes could spread disease, which is a legitimate concern because it's happened with other reptiles. It is difficult to find homes for research animals in general and near impossible in the case of snakes, especially venomous ones. But I think that few scientists even try to find a home for their 'experimental subjects' once they're no longer of use."[89]

Amarello and Smith live with four rattlesnakes: "During graduate school, we encountered a pregnant Arizona black and brought her into the lab where she gave birth. I was given a couple babies from a friend as well. When my doctoral project was scrapped, I tried to find places where the snakes could be homed since they could not be released into the wild. I was able to place three with other educators and then kept the remaining four, whom we use for public education. They are four male Arizona black rattlesnakes, Cash, Kai, Snow White, and Hicks, who were born in captivity in September 2009. I haven't measured them in ages, but I'd guess they're around eighteen to twenty-four inches long. It's great to be able to show people a rattlesnake in a controlled, unfrightening experience so they can see that they are not out to get us. But, I really wish I could release all of mine into the wild."[90]

Melissa still considers herself a scientist, but not in the old mold. The science she conducts today is "for the snakes, not to prove any human thing, but to help us understand how we can better co-exist with these misunderstood wonderful animals. That is why we founded our nonprofit to educate and teach people how to do just that. There is a huge need for positive messaging about snakes. I had seen firsthand how negative attitudes can stifle, and even backfire, conservation efforts and that is what I want to change."[91]

D. H. Lawrence wrote a poem, "Snake," about an encounter he had with the mysterious serpent. His lyrical sketch reflects the conflicting emotions that rattlesnakes evoke, and, similar to many herpetophilic scientists, he questions human, not snake, motivations.

> A snake came to my water-trough
> On a hot, hot day, and I in pyjamas for the heat,
> To drink there.
> In the deep, strange-scented shade of the great dark carob-tree
> I came down the steps with my pitcher
> And must wait, must stand and wait, for there he was at the trough before me.

The writer stood there, watching as the snake

> rested his throat upon the stone bottom,
> And where the water had dripped from the tap, in a small clearness,
> He sipped with his straight mouth,

> Softly drank through his straight gums, into his slack long body,
> Silently.

But Lawrence's wonderment turned to worry:

> And voices in me said, If you were a man
> You would take a stick and break him now, and finish him off.

Then, another voice played in his head, this one retaining the poet's initial sentiment,

> I confess how I liked him,
> How glad I was he had come like a guest in quiet, to drink at my water-trough,
> Again, inner conflict sparked anew,
> Was it cowardice, that I dared not kill him? Was it perversity, that I longed to talk to him? Was it humility, to feel so honoured?

> Eventually, the stern voices of the collective won out.
> I picked up a clumsy log
> And threw it at the water-trough with a clatter.

At once, Lawrence was gripped with sadness:

> And immediately I regretted it.
> I thought how paltry, how vulgar, what a mean act!
> I despised myself and the voices of my accursed human education.

Soon, Lawrence retired to the house. He closes the poem with these words,

> And so, I missed my chance with one of the lords
> Of life.
> And I have something to expiate:
> A pettiness.[92]

The *scientia amabilis* of old that neurosciences has re-ignited echoes Lawrence by entreating scientist and citizen alike to expiate the "pettiness" of human privilege and take a chance with getting to know our reptile kin.

6
Pumas: Psychological Trauma

Everybody said the same thing: "There's something different about her." The tawny power of her walk was spellbinding. She moved like she owned the world, and when she looked at you—if you were ever that lucky—her stare held you captive until she decided to let you go. But with her daughters, she was completely different. Any trace of suspicion disappeared in the glow of love. It was easy to see that the children were her life—literally, for it was because of them she died.

The Wyoming day began like many others. Sunlight pierced through brilliant blue skies onto sparkling snow. It was a stupendous morning. Two sleepy bundles lay nestled in the grass. At the sound of their mother's murmurs, the cubs awoke. She had been up for hours and was just returning. After a tender reunion, the constellation of three rose and left. It was never safe to stay in one place too long and besides, there was food to be found.

The two sisters walked single file following their mother, eager for novelty, enthralled with a world that had yet to threaten. Suddenly, they heard her hiss a warning, "Run and hide!" Her voice was strange, harsh and dry. They immediately obeyed and crouched behind a bush. In a flash the world exploded with angry screams and the sound of rending flesh. The two kittens huddled closer together, cowering. Another voice, deeper and frightening, tangled with their mother's. More screams. An ear-splitting wail rose to the sky, then arced downward into silence.

After a few minutes, the girls ventured out. They smelled a familiar acrid odor. It was blood. Their mother lay still, almost posed, in a stretch of red. Save for the lonely sound of a few tumbling rocks and the cheerful chickadee's call, it was silent. The puma did not move. She was gone, and the kittens were alone.

So goes a page from the book of puma lives. The three pumas, kittens F99 and F75 and their mother, F51, are part of the Teton Cougar Project (TCP), a long-term study by Panthera, a nonprofit organization dedicated to the research and conservation of big cats, most species of which are sliding toward extinction in one way or another. It's hard to believe that the world's most powerful land predators are in trouble. After all, what could be mightier than an African lion who can pull down an African buffalo five times his size? The answer, as with all other endangered species, should no longer surprise us: modern humans.

Five centuries of hunting and habitat loss have decimated the kings and queens of savannah and jungle. Lions are predicted to become ecologically extinct; traces of their somatic forms may linger, but they will have lost the ability to function as a thriving society. Only 450 Siberian tigers remain in

F51. Photo credit: Mark Elbroch.

the wild. The North American puma (*Puma concolor*) and its sympatric companion, the jaguar (*Panthera onca*), may be more numerous, but overall, they have not fared much better than their more exotic cousins.

As reflected in the more than eighty common names and multiple subspecies, pumas once roamed unhindered across North America from ocean to ocean and down through Central and South America. Today, however, like the grizzly, pumas occupy only a fraction of their historic habitat. Cougars are systematically killed not only with governmental approval, but also with its encouragement and incentives. The puma's complete annihilation has so far been only narrowly averted by the requirement of hunting permits. Across sixteen states between 1992 and 2014 there were 151,863 reported cougars killed, 95 percent by guns.[1] Most populations have been extirpated. Florida has only one breeding group left, with a frighteningly low number of less than one hundred. These cougars, along with their California counterparts, are experiencing alarming losses of genetic diversity, a bottlenecking due to burgeoning human development.[2] Yet despite their diminished numbers and shrinking homelands, pumas are afforded little of the reverence that their African and Asian counterparts receive.

While Cecil the African lion and his kin are celebrated as trophies, outside an occasional film showcasing the puma (also called the "American lion," mountain lion, and cougar), the species is relegated to the rank of pest. This judgment was seeded as early as the 1500s with the Jesuit priests' offers of bounties for killing pumas and Theodore Roosevelt's damning condemnation.[3] In the same tradition, Bruce Babbitt, U.S. Secretary of the Interior from 1993 through 2001, once recounted, "I grew up in a ranching family and we've been very much a part of this history of eradicating predators. The paradigm of ranching was to clean the rangeland of all threats not only to the livestock but to the grazing base and that meant eliminating [carnivores, and] creating a silent landscape." Even though pumas' recategorization as a game or trophy species in most states has saved the species from extinction so far, the big cat is still not tolerated. This explains why hardly an eye bats when wildlife officials and ranchers hunt down cougars accused of trespassing or killing livestock. Such intolerance makes for hair-trigger agency responses to sightings that leave no time for judge or jury. Government-sanctioned and financially underwritten predator-control programs enlist steel traps, leg-hold traps, and snares as well as hounds and guns.

Most scientists admit that wildlife laws of zero tolerance are based on hype, and are overblown and oversold as a supposed need to control livestock depredation and mitigate dangers to human life. As in the case of other carnivores—sharks, orcas, crocodiles, and so on—the number of cougar-caused human deaths is minuscule. Defenders of the species point out that the risks associated with dog bites, hiking, camping, driving a motor vehicle, and the mild-mannered sport of golf are exponentially greater. But facts fall on deaf ears. Despite exonerating statistics that roll out year after year, the tawny cat continues to be hunted with cold-blooded efficiency and resolve.

Since the colonization of the Americas, the frontiers have been deemed too small to share with native cats. Pumas are big, as are their needs for space, reflected in the expanses of this continent that they historically inhabited. They can reach up to 160 pounds and over six feet in length, a third of which is taken up by a nuanced rope of a tail.[4] They are lean leapers of grace and power. The size of puma home areas varies with the amount of food, water, and terrain as well as "intrinsic factors" that include gender and body size. An average estimate of home ranges for males is about 100–150 square miles, and females about 50–100 square miles.[5]

Home range is an apt description, because cougars regard the entire area in which they roam as home, which, unlike a territory, is not defended. Males, and now it is thought, females, create scrapes, which are sticks and leaves scratched up into a small pile and sometimes topped with urine or scat. Such cougar blazes are polite communiqués demarcating property and boundaries. In a similar fashion, both genders will cover their kills, even if it is symbolic. Harley Shaw, a wildlife biologist with the state of Arizona, once came across a kill perched atop a naked boulder upon which a cougar had daintily placed a single twig. This points to a cougar philosophy of studied avoidance, rather than defensive aggression.

Second only to their reputation as masterful hunters is the puma's legendary elusive nature. As one of the most difficult wildlife species to study, researchers traditionally hire houndsmen and their dogs to track and tree the big cat. Shaw studied cougars for more than three decades, employing lion houndsmen to track and locate pumas for research purposes. He came to know the inner workings of the puma hunter culture firsthand, the center of which is the hound, bred and trained to find mountain lions. An experienced team of canines and *Homo sapiens* makes a formidable predator of the continent's largest feline.

For lion hunters, trained hounds are required equipment. Assembling and training an efficient set of dogs is no easy task, according to Shaw. Grooming hounds for big cat pursuit is a far cry from ordinary dog training: "You do not train them to do something—to retrieve, to sit, lie down, and so on. You train to eliminate undesirable behavior while retaining the desirable."[6] Among the desirable qualities is focusing on the prey at hand, a mountain lion, not dog-palatable deer and rabbits, whom houndsmen refer to as "trash."

There are many ways to make a hound lioncentric. As Shaw explains: "Hound advertisements abound with solutions and scents of unwanted game to be used in negative conditioning of trashy dogs, and many exotic techniques for applying these scents have been developed. A common tool is to cage the errant canine with deer scent, or even deer parts, in a barrel rigged much like those once used for lottery drawings, and crank away. The wild ride with the scent is supposed to do the trick. I've also heard of injecting a nausea-inducing drug, then releasing the dog on a trash trail. Collars designed to shock the dog during the unwanted act are considered to be one of the best modern tools for training particularly stubborn hounds."[7]

Shaw describes the basics of a hunt once the dogs are trained and tuned up to go: "A hunter's first responsibility . . . is to determine the direction of travel of the cat. On snow, this is not a problem; on hard dry ground, one can spend hours trailing in the wrong direction before crossing a visible track . . . once the dogs are lined out on a track, if it is a fresh one and if the dogs are reliable, the hunter can relax and just tag along . . . The idea, of course, is to trail the lion where it walked the night before and stay on the trail until the cat is jumped. Lions normally bed up in thick brush or boulders or under overhanging ledges in midday. As a result, daytime tracking, if you are on the right end of the track, and if the track itself isn't too long, should lead you to the lion."[8]

Cougar hound hunting is an old tradition and remains a viable sport and profession for hire by government agencies. Giles Goswick, born in 1888, is a legend among carnivore scientists and hunters: from the age of thirteen, when he shot his first cougar, and over the next sixty-four years, he killed 960 cougars, most within the confines of a single Arizona county, Yavapai. Others, like Ben Lilly who lived in the mid- and late 1800s, utilized their hounds for other prey, including bears, coyotes, and other predators. The Paul Bunyan of carnivore killing, Lilly racked up over a thousand cougars in addition to the staggering numbers of other "vermin" that he killed.[9] Then there were Jay C. Bruce and C. W. Ledshaw, two lion hunters hired by the State of California, who killed 922 cougars in less than three decades.[10]

When pioneers made their way across the continent, no one questioned the mass slaughter of bears, coyotes, wolves, and cougars until the John Muirs and Henry David Thoreaus protested the wholesale decimation of wildlife and demanded legal intervention. The government responded, but since ranching and grazing interests were so influential, among the first foundation stones of federal policy was conservation-based killing. In a 1966 interview, old-time lion hunter Goswick reflected the weird-speak of contradictory missions and meaning that persist today in wildlife agencies and policy. The hunter was reported as saying that he strongly believes in the balance of nature and voiced opposition to categorizing lions as game animals with a specific season. This, he asserted, would soon drive the cat toward extermination. "We're not out to exterminate the lions. We want to keep them under control." Goswick lamented the fact that in "those days"—earlier times—"there was something to hunt" and expressed concern over what he saw as a scarcity of game.[11]

It is incredible that a man as practical as Goswick failed to understand the link between the obliterating hunts in which he engaged and the scarcity of game at the time of that interview. Nonetheless, neither numbers nor age slowed the hunter. At age seventy-seven, he continued to drain the Southwest of its carnivores. Two months before the interview, Goswick, along with his son and grandson, had killed ten mountain lions.

Today hounds are still used to hunt puma, but with a digital twist. Both researchers and hunters have made the switch to radio telemetry. Private hunters and those paid by state and federal wildlife agencies are taking advantage of intimacy-at-a-distance technologies by radio-collaring their dogs, who are then let loose to track their prey. Electronic tracking is far more comfortable than braving the elements. The dogs do almost all of the work while the hunter stays inside by the computer with a hot cup of coffee. When radio data indicate that a cougar has been treed, the hunter calls his client, then leisurely drives out to the spot where the dogs are keeping the puma at bay. Lodged in the limbs of a tree, pumas make easy targets for a hunter who can take her or his time, light up a cigarette, then amble up for a safely aimed shot. Biologists hire houndsmen to do the same thing, but without the intent to kill.

It was these new data sources combined with time-tested, sharp woodsmen skills that enabled Panthera wildlife biologist Mark Elbroch to piece together the story of misadventure and untimely death of the mother puma F51: "The tracks in the snow were a day old and completely intact so we were able to get a pretty good picture of what happened. M85, a full-grown male cougar, had been at his kill for two days at the base of spectacular red cliffs. It was then that F51 and her two kittens headed down a narrow slot in the cliffs above M85's position. The draw didn't have any snow, and there were lots of loose rocks and the three probably made quite a bit of noise. Plus, we don't know for sure, but the kittens, as young as they were, might have been playing around. They may have even been in front of their mother."[12] It was clear who the killer of F51 was. But less obvious was why the tangle had happened at all, until a little more backstory yielded some inferential context.

Several months earlier, before F51 died, a hunter had shot and killed M29, the resident male and father of her kittens. Because nature and mountain lions abhor a vacuum, M85 began encroaching on M29's now-emptied home range. When M85 encountered F51, he was outside his usual area. When the male and female encountered each other, they were within the perimeter of M29's home range.

The first thought that came to biologists' minds was that M85 had killed F51 in an attempt to kill her kittens. But the snow said otherwise: "It was F51 who was going fast and furious, not the male. F51's tracks showed that she started to pick up speed. M85's tracks showed that he literally walked up to where the encounter occurred. Taking both sets of tracks together, it's clear that F51 was the one who initiated engagement. Whatever the other details of the scenario, whether the kittens were ahead of her or behind, the pair met in a storm of claws and fury, packing the snow as they wrestled. They slid down the hill again and again, rolling sixty to seventy feet, and leaving behind great tufts of fur. In the last great tumble, the pair slammed into a young fir tree, snapping off its lower branches."[13]

M85's non-aggression agreed with the data that Panthera had been collecting over the years. Well before the present kittens, F99 and F75, were twinkles in their parents' eyes, F51 had had a litter with M29's predecessor, M21. At the same time, M21 had been stepping out with another female, F61. In what seems to have been happy polygamy, the two females hung out together and continued doing so even after they both bore litters of bouncing kittens sired by M21.[14] Sadly, M21's fatherhood was short-lived, because he died from ingesting rodenticide in 2012. It is not a nice way to die.

Enter M29, the future father of F99 and F75. Foreshadowing the same pattern in the future when M85 would expand into M29's home range after his death, M29 took up residency in the void created by M21's passing. Although evolutionary creed would predict that M29 would burst onto the scene bent on exterminating his predecessor's progeny, the three pumas (two female adults, F61 and F51, and an adult male, M29) formed a tightly knit trio that spent time together, which included the sharing of kills. In every way they operated as a family unit, even to the point where M29 mated with the females, and with nary a sideways glance at their kittens.[15] He and F51 were to become parents of two batches of beautiful kittens. Given this history and circumstances, Elbroch notes that

> it didn't make sense that M85 would attack F51. M85 was in the height of breeding season and had been carousing with females the last few weeks. It made much more sense that F51 would be a potential mate rather than an adversary or threat. This interpretation is consistent with what we have been seeing for years. We have never seen any aggression between males or between males and females. Time after time, when males meet up, they typically avoid each other or "turn the other cheek." Out of 78 kittens in our study, only two were killed by another puma . . .

> This in turn leads us to another question: is it common that clashing cougars results in the death of one or the other? According to 14 years of research, the answer is yes and no. Yes, in that we have documented cougars killing other cougars . . . but, no because it's also very rare . . . Why a male would kill an adult female is . . . difficult to explain with biology. We expect it had something to do with F51 defending her kittens.[16]

When an intraspecific killing happens in the animal kingdom, academic pulses quicken. The reigning theory for nigh on forty years has been that male carnivores and other species, including primates, go out of their way, Richard III style, to kill unrelated young to clear the genetic pathway for their own spawn. By killing the progeny of their rivals, the logic goes, males gain the opportunity to mate with females who return to estrous in the absence of cubs. Evolutionary biologists interpret such naked villainy as a strategy employed by males to increase reproductive benefits by knocking out competition—what biologists refer to as increasing inclusive fitness—and/or to use the progeny as a source of food.[17]

Murder, particularly infanticide, is a troubling issue for scientists and laypeople alike, as it seems to tear at the social-moral fabric that binds a culture and species. Primatologist Craig Stanford writes, "Of all the behaviors primatologists have observed, none has fascinated or appalled us as much as infanticide. When Jane Goodall reported a mother-daughter team among her Gombe chimpanzees (the infamous Pom and Passion) preying on the infants of other families, we were shocked by yet another brutal behavior from a species so like ourselves."[18]

Shocked—and piqued. The my-genes-made-me-do-it theory is a sacred and sensitive cow. Those who have questioned the data and prevalence of infanticide have come under heavy academic artillery fire.[19] When Anne Innis Dagg published a detailed analysis showing the gaps between theory and data in lion infanticide studies, critics who view infanticide as a compelling evolutionary adaptation reacted angrily. Yet even they admit that "direct observations [of infanticide] are difficult to obtain and most of the mortality has been inferred from demographic data" as well as that "the total number of direct observations of infanticide may be small." They also insist that discrepancies between their conclusions and those of others who disagree stem from insufficient study and a lack of understanding of the "great infusion of evolutionary theory into the behavioral sciences in the 1970s."[20]

But other researchers from various fields are also beginning to question the evidence behind the "kill-to-conquer" theory. For example, incidents of intraspecific chimpanzee violence (which are often pointed to in support of the theory) when reexamined appear related, at least in part, to human-caused social disruptions through the manipulation of food resources in study areas, including Gombe Stream National Park. Ostensibly in protected locations, the study areas are nonetheless situated where poaching and habitat destruction are widespread.

Rutgers anthropologist Brian Ferguson, whose work has focused on the study of war and its causes, found during his investigations that when "considering certain and very likely killings . . . there are 23 consistent with maximizing inclusive fitness and 25 that go against inclusive fitness," which makes any definitive identification of causality "pretty much of a wash."[21] In other words, there is no clear case to support the biological basis for war, and the data purported to link infanticide and other intraspecific violence are insufficiently consistent. What is known with certitude is that chimpanzees are susceptible to psychological trauma and that all populations used to study lethal conflict have sustained severe violence by humans.[22] Even the most parsimonious interpretation argues for trauma as a significant factor in shaping wildlife lives and minds.

The primary bones of contention in the infanticide debate seem to be prevalence and purpose of the act. The prevalence, as Ferguson and others have pointed out, is variable, and the circumstances specific. The data do not support universal rules for when, who, and why a lion, chimpanzee, or other species will kill one of its own. Furthermore, the theory that seeks to establish the demon male model is largely based on the belief that behaviors are more geared to evolutionary advantage rather than to particular situations playing out in real time—which in itself is difficult to prove, particularly now that neuroscience and epigenetics have thrown their hats into the explanatory ring.

Epigenetics has already made significant inroads where humans are concerned. On the heels of multiple lines of research, epigenetics dealt the final coup that dislodged behaviorism from its hegemonic position by "forc[ing] the recasting of a centuries-long philosophical debate." Cartesian splits between nature/nurture, mind/body, and so on have been transformed from agonistic pairs to cooperative partners, so instead we see that "there is a dynamic interplay between genes and experience, a clearly delineated and biochemically driven mechanistic interface between nature and nurture." In this way, "the emerging field of neuroepigenetics has necessitated the reformulation of the fundamental existential question of nature versus nurture."[23] When we take into consideration contributions from neuroscience toward a better understanding of animal sentience, pumas and other animals shed their image as biological automata to become thinking, feeling, and critically, independent agents. How lions and chimpanzees act is no longer understood as an expression of autocratic genealogical mandates but rather as genetics experienced in an environmental milieu brokered by a social sentient being. As a result, evolution alone as a reason for infanticide loses quite a bit of its explanatory thunder.

So much uncertainty clouds the inferences made surrounding infanticide, yet the scientific discourse remains acrimonious, and defenders, vociferous. These conditions suggest that something more than angst fuels the debate. There is also concern for how erroneous scientific conclusions about nonhumans may be used and interpreted in legal and ethical venues to make incorrect inferences about human behavior.[24] Dagg cautions, "What needs to be explained is not the reluctance to accept distasteful theory, but the willingness to promulgate unfounded theory . . . No one would deny the existence of many genetic sources of human behavior. However, sociobiology treads on dangerous ground when it incorrectly gives unwarranted prominence to the biological basis of human behavior."[25]

Academic disputes aside, can the topic of intraspecific killing be approached in a way that yields insights into the cases of F51 and others? In many ways, neuroscience can contribute, if not to provide conclusive answers, then to infuse equanimity into a polarized debate. By looking more closely at puma social and ethical development through the lens of neuropsychology, a more insightful perspective can be gained.

Much of what was learned from Bowlby and the bears also applies to cougars. Although wild felines do not employ hermetically sealed hibernation dens like those that nurture grizzly minds, female mountain lions like F51 do give their offspring shelter from the storm with thoughtfully located and crafted dens. Mothers then shift their brood from den to den within their home ranges. Like nomadic homes, their nesting areas are not permanent, but the shelter of the maternal tent is.

As the kittens are weaned and continue to mature, the mother familiarizes them with the nooks and crannies of home. This intense, protective mothering acts as broker between what young pumas experience externally and what is taking shape internally. The better the fit between external realities and internal expectations, the better a young puma can meet and match the demands of a sometimes unforgiving environment.[26]

At about two months, young pumas are fully weaned, but they remain with their mothers for another sixteen months, a good sixth of an average puma lifespan. During this time, young brains and minds are neurobiologically infused with the rudiments of puma culture, to be passed down generation after generation.

Traditionally it has been assumed that pumas existed as a "currency of one," that they lived as solitary wanderers and met their conspecifics solely for mating or rearing cubs. According to this view, the developmental attachment patterns of pumas derive almost exclusively from the mother. Other intraspecific relationships were considered distant at best and lethally conflictive in the extreme. But once again, technology has brought new revelations to what has hitherto been obscured by the puma's covert arts.

To catch cougars unawares, proximally undisturbed and unhandled, Elbroch and his fellow researchers set up motion-sensitive trail cameras and lights along known wildlife paths or at places where cougars have made kills. When a puma wanders into the path, the camera is triggered to record a photo or video. Working by day and night, the cameras are perfect for recording the nocturnal ways of the big cat on the prowl.

Three kittens of F51. Photo credit: Mark Elbroch.

Elbroch hoped that the cameras would shed light on the types of cougar behavior that researchers either cannot observe at all or see in only fleeting glimpses. But what the wildlife biologist didn't suspect was that the cameras would show "up to nine pumas on a kill at the same time comprised of multiple families: moms, sibs of this year, and an independent male."[27] Though pumas *en famille* goes against the assumptions of most mainstream scientists, the recorded data are corroborated by old-time lion hunters who "were sure that mature toms and females with kittens fed on the same kills."[28] F51 herself was filmed feeding with M21 and then months later with M29—both males.

To get a panoramic view of cougar society in which the natal families are embedded, Panthera scientists superimposed radio telemetry on home range data, then performed a network analysis that created a kind of paint-by-numbers sketch of cougar sociality. Similar to the Amarello-Smith rattlesnake study, the cougar analysis quantified who is interacting with whom and how often across space and time. This social network analysis lined up nicely with male home ranges, and showed that there is quite a bit of interaction among females.[29] In fact, it is not uncommon for clusters of four or five females to interact within a given male's home range.

The Wyoming study has made significant strides in transforming the image of pumas as loner killers itching for a fight into one of a species that operates in family units. American cats are a lot more family-oriented than previously thought. As neuropsychologist Narvaez asserts, such "affiliative capabilities form the grounding for what Darwin called the 'moral sense'" in humans and in other animals.[30] The Panthera studies also expand the conventional scientific aperture from a narrow focus on stereotypical paired behaviors (mating, fighting, feeding) to a broader collective view of puma society. The species' social floor plan may differ from that of orca pods and Africa lion prides, but puma society has its own architectural complexity that informs young puma brains and minds in much the same way that Urie Bronfenbrenner once envisioned human infant development.

Bronfenbrenner, a Russian-born psychologist, was instrumental in bringing attention to the effects of the environment on a child's mind—in stark contrast to the then-dominant theory based on biological determinism. His idea that linked child behavior with social-ecological experience contributed to the founding of the Head Start program in the 1960s under President Lyndon B. Johnson.

Bronfenbrenner's model embeds infant cognitive, affective, and moral development in a dynamic matrix of nested, interacting systems, ranging from the externals of climate and ecology, to the species-specific of genes and culture, to the internals of physiology and the autonomic nervous system. Like a never-ending Möbius loop, what an infant experiences on the outside in his or her environment interacts epigenetically and neurobiologically on the inside to inform individual and group morality, which in turn feeds back into the socioecological surroundings.

Unlike modern urbanized humans living in James Hillman's "anorexic," spare landscapes, Quechua, orca, and puma brains and minds are ecologically engaged and in dialogue with nature in its entirety and are reflective of our ancestral small-band hunter-gatherer society. These are societies characterized by "cooperative, collaborative orientation," while "at the same time there is high autonomy" that cultivates a morality informed by "small egos but 'large' selves, the common self . . . [and] includes a sense of empathetic concern for family and community."[31] A prosocial milieu may assume various shapes. At one end of the *Panthera* genus is the spatially close African lion pride with its synchronized, choreographed hunting and rearing, and somewhere further along the spectrum, the more diffuse, but coordinated, mosaic of mountain lion home ranges linked via relational transactions. Each model proffers its own rules of moral engagement that have coevolved with the environment.

Mark Elbroch describes an example of puma etiquette: "Mountain lion table manners require that they eat at a suitable distance from one another. If the carcass is small, this means that mountain lions must take turns eating at the carcass, but if it is large, like an adult elk or moose, two and sometimes three mountain lions may feed at the same time. These rules are concrete. They cannot be broken. Any mountain lion that violates the rules of respectful distance between feeding mountain lions or moves too close to a mountain lion that is eating, will elicit hissing, swatting, and/or lunging. Feeding mountain lions react ferociously and with blinding speed."[32]

In the fast-food-pay-your-own-tab lunches of urban and suburban life today, it is easy to overlook the profound significance of breaking bread together. To share in a coveted kill is to share life, and it shows a willingness to provide for another even though it diminishes one's own portion. This reflects an understanding of prosocial interconnection and its value. That pumas share food with other adults argues for an ethic of care that supersedes the selfish gene.

After they leave their mothers, pumas continue on their rapid ascent to sexual maturity. At two years, just six months after the average dispersal age, young pumas enter early adulthood. A number of theories have been proposed for what prompts dispersal—namely that the mother comes into estrous, the mother is about to give birth, or there are internecine disputes. Similar to grizzly families, the prevalent pattern of good-byes typically seems to coincide with the incipient birth of a new litter. A month before she is due, the mother generally draws the line and sends her teens on so she can gather strength and resources to focus on the next generation.[33] This was the case for our friends, the mother pumas F51 and F61, as the thread of their biographies demonstrates.

In June 2012, the puma mothers F51 and F61 separated with their litters. Instead of sticking with the mother, one of F51's kittens, F88, left in tow with her allomom, F61. Cherished and coveted puma young shared across families! Perhaps the rationale derived from a difference in parenting styles. F61, who was noted for her more old-fashioned mothering, might have been deemed more capable of caring for an extended brood. This may have taken the pressure off F51, who was hugely loving, but less adept at providing material comforts (she was what Elbroch called smilingly, more love and peace, "like a hippy mom").[34]

Eventually all four kittens left their respective biological or adoptive mothers at the age of fourteen months. This was a tad unusual in that it was four months sooner than when young adults typically disperse. But consistent with what Elbroch has observed, the young pumas' dispersal occurred a month before both females gave birth to litters fathered by M29.

Again, like grizzlies, some mothers show ambivalent feelings about their brood's departure from the natal fold. F61's kittens, M80 and F96, remained with her until she kicked them out at nineteen months. Trail videos capture the growing, mutually shared feeling that it was time to cut the apron strings: "M80, as he grew to outweigh his mother[,] increasingly violated [house] rules. He would slowly slink into close proximity with his mother while she was feeding, pausing at regular intervals as if testing the waters for her response. When he had slithered in close enough, he'd throw his bulky hind end into a position separating his mother from her food, and then he'd cringe in place and await her wrath. Depending on her mood and how much she'd eaten, she might just walk away, or move off after a few perfunctory swats to remind M80 that he was living in her house. If she vocalized and swatted during his approach, he knew not to push her any further and he'd retreat to await his turn."[35] Seven months after leaving the nest, M80 was killed, and F96 had established her own home range.

In summary, puma kittens are born into and nurtured via a secure attachment and remain part of this natal molecule until just short of sexual maturity. Consistent with Bronfenbrenner's nested view, it appears that there is another social valence in which a mother-kitten constellation may participate, a kind of overlapping crèche with an additional allomother. Puma and grizzly relationships function as a spatially diffuse social network, but as trail cameras reveal, cougar relationships are sticky. Teens may hang on to the family, sharing meals with mum, dad, and perhaps a new litter, and leave, as M80's and F61's relationship showed, only when exigencies of an impending new litter require it. The glue that makes these bonds sticky is not just the practical need to perpetuate one's genes—it is emotion. F51's fight to preemptively safeguard her kittens says it all.

With a sketch in hand of a secure, presumably historically normative cougar family life and development, we can now revisit the question of intraspecific killing. To do so requires looking to some additional ongoing studies with some less than salubrious implications. For while Panthera biologists have been conducting field work in the Rockies, researchers in California and Washington have found pumas being surprisingly visible in violation of their legendary cryptic nature.

Historically, puma sightings are rare. They typically avoid roads and other areas that make them vulnerable to humans. But pumas have had to adapt to the progressive dissection of their habitat by encroaching humans, including roads, development, and ATVs that go anywhere they want. With no other choice, pumas are now starting to show up in unprecedented realms, and in broad daylight. The University of California, Berkeley, campus has its own resident cougar family.[36] In southern California, a National Geographic photographer caught a lone male cougar wandering through densely populated hills under the iconic HOLLYWOOD sign.[37] Like many wildlife celebrities, the young puma has been given his own Facebook and Twitter accounts under his name, P22, which evokes an unfortunate association with the German Walther pistol.[38]

P22 lives in what many call the "world's most urban park," Griffith Park in Los Angeles. Genetic tests show that he is new in town and related to pumas living to the west in the Santa Monica Mountains. As the sole lion in the Park, P22 was assumed to have traveled the distance on foot to get to his new digs. Genetic tests have confirmed this hypothesis.

To help orient less geographically savvy readers, P22 dispersed from the peaks of Topanga Park overlooking the Pacific Ocean to the southeast, continuing toward Griffith Park by crossing one of the busiest manmade corridors in the nation, Interstate 405. Nearly half a million vehicles flow up and down the steel river each day. Traffic jams are so horrendous that locals claim the freeway got its name from commuters crawling at only four or five miles per hour.

After crossing the multilane freeway, the young puma wove his way through such elite neighborhoods as Bel Air and Beverly Hills. Although the traffic might have been more diffuse in suburbia, the padded wealth of lawns and gated communities are nonetheless riddled with roads and cars. To reach Griffith Park, the puma had to, once again, brave a freeway gauntlet, this time, Highway 101. His hazardous journey, seventeen miles as the crow flies, took him twenty-six miles on foot.

Once spotted, park officials tracked, caught, and collared P22. He was treated for notoedric mange, a skin condition caused by mites that has become epidemic among wildlife in the area. Between 2002 and 2006, over half of the radio-collared bobcats in the Santa Monica Mountains National Recreation Area died from the disease. Mange is suspected to have originated with domesticated dogs and cats. While not usually a threat, mange continues to spread and it can be deadly for those with compromised health. P22's bloodwork showed that he had consumed diphacinone and chlorophacinone compounds, part of an anticoagulating rodenticide that reduces immunological resilience.[39]

P22 continues to be studied as part of the Griffith Park Connectivity Project, but biologists do not consider his cross-country journey a success. He may have survived, but "he remains hemmed in by multiple freeways and has no opportunities for reproduction."[40] True, no man or lion is an island, but anyone familiar with L.A. traffic and its urban terrain would consider his feat something worthy of the Guinness Book of Records. It appears that P22 is getting used to this new human-dominated lifestyle. One day, he was discovered sleeping in the crawl space of a residential house. "He was just laying there," the homeowner told reporters, "trying to snooze, completely just like, we woke him from a nap."[41]

Meanwhile, in Washington State, other young male pumas are venturing deep into the bastions of humanity.[42] The unprecedented incursions are usually attributed to exploding populations or overhabituation—the loss of what is regarded as a healthy, natural fear of humans. But scientists have noted a correlation between these areas of incursion and areas where adult pumas are hunted selectively to keep the species in check. In the southwest United States, where bighorn sheep prized by hunters live, adult female pumas are killed under the assumption that males will then leave. The absence of big adults can, however, create a territorial void that invites in younger males.[43] Combining new methods of tracking and trail cameras, and the neuroscience of animal sentience, there is yet another factor that brings clarity to the mystery of pumas acting in very un-puma-like ways.

From the famed but ethically bankrupt experiments of Harry Harlow, to the excruciating testimonies of refugees and concentration camp survivors, science and history are replete with the mind-shattering and life-altering impacts of psychological trauma. For carnivores, the story is eerily similar. With drastic losses of habitat, a constant threat from hunters, high mortality, and unreliable food sources, life for the average carnivore has changed dramatically and rapidly from historic norms. Under highly stressful physical or emotional conditions (food deprivation, decreased habitat, loss of one's mother, social disruption), species-normative brain processes are compromised.[44] What goes around on the outside, comes around on the inside. Each unusual change in the environment telegraphs directly into the brain and body, altering the organism's inner blueprint. These neuroepigenetic changes then are expressed as variations in personality, stress regulation, and immunological resilience.[45] The result is a puma who is not quite a puma.

When puma mothers like F51 are killed, their orphaned kittens are prematurely weaned and deprived of the nutrition and care that cultivate a healthy lion mind. Unprepared for harsh climates and terrains, unprotected kittens frequently become prey to other carnivores or simply starve to death. If they do survive, they must learn on their own how to deal with an increasingly complex and hostile environment. The constant threat and impact of routine hunting, which is the major cause of large carnivore mortality, as well as restricted access to food, act as chronic stressors. Studies on black bears in the United States and brown bears in Sweden found elevated heart rates and depressed heart rate variability, both of which indicate stress and anxiety associated with human settlement proximity and presence. Researchers came to the same conclusion: humans have created "landscapes of fear."[46] Coupled with earlier traumas and the absence of elder guidance, both of which compromise normal development of social and emotional intelligence, the addition of chronic stress may trigger risk-taking behaviors and aggressive and fearful mental states. The pattern correlates with F51's history and experience.

In 2011, when Panthera biologists began tracking F51's movements, the future seemed bright for the female puma. She gave birth to three litters in three years and her first litter in 2011 was a prodigious bounty of five kittens. (Typically, in Wyoming, there are three to a litter.) The joy was short-lived. Only three of the five kittens (M36, F59, and F88) survived long enough to go it alone at fourteen months of age, and F59 was killed just two months later. The late-year births followed by an early dispersal were risky. Wyoming climes transition quickly from nurturing summer to harsh, lean cold. As the air chills, many prey hibernate, making warm-blooded meals increasingly difficult to find.

Then, in the fall of 2012, F51 gave birth to a modest three kittens. Within weeks, however, she once again suffered loss—wolves killed two of her new brood. F51 fled with her remaining kitten F70, whom researchers named Lucky. Mother and kitten ran two miles nonstop away from the site of the attack. They found shelter in a cave under a cliff and stayed there eight consecutive days without food—quite a stretch and lots of missed opportunities that are particularly critical in winter. In a study on kill and consumption rates of pumas in Colorado, Patagonia, and California, Elbroch and colleagues calculated that pumas move on average 7.6 and 11.8 km a day, although there is quite a lot of associated variance. Kill rates are fairly consistent across geographic locale, gender, and body size, although females with kittens showed a marked increase over others. Although feeding frequency varies with availability as well as the ages and numbers of dependent young, a female with kittens needs to eat the equivalent of about one good-sized mule deer once a week.[47] Significantly, there is also a high rate of kill abandonment. The reasons for this are undetermined but may include some sort of foraging strategy or kleptoparasitism.[48]

During this period of cliff retreat, F51 circled back twice to the place where her kittens had been killed. For the next three weeks, she repeated this pattern, and when she did hunt and make a kill, she would not go outside a three-hundred-meter radius from the cliff hideaway. Finally, after a full month, they left. Then, as Mark Elbroch put it, "the unthinkable happened." F51 mated with M29 and separated from Lucky when the young puma was only nine months old. Two months later, a hunter killed Lucky.

F51 gave birth to a third litter of four bouncing balls of spotted fluffy fur that included F99 and her sister, F75. It was another late-season birth, but F51 "had chosen a wonderful den." When Panthera researchers caught and radio-collared them at about five to seven weeks of age, the fifteen-minute health check indicated that all were in good form. The male M46 was tiny, though, weighing a pound less than his sisters. It was a vicious winter with temperatures as low as −35° F, and in December he got wet and died from exposure. A month later in January, a wolf killed one female kitten, and in March, F51 was killed by M85, leaving kittens F99 and F75 alone to fend for themselves. F59 and M36 dispersed upon their mother's death. F59 was killed shortly thereafter, and F88 (who was adopted by F61) dispersed successfully. Of F51's twelve total offspring, only two could have survived to adulthood, and F51's life was prematurely truncated.

Details of P22's past provide additional evidence supporting the theory of puma psychological trauma and collapse. The young male's tortuous journey through Bladerunner Land has a more sordid twist than a simple James

F51 and kitten. Photo credit: Mark Elbroch.

Dean saga about a rebel teen seeking to make his mark in life.[49] His father, P01, killed his mate P02 and two others, then mated with his daughter. There have been other killings and inbreeding in the case of other cougars occupying the Santa Monica Mountains.[50]

Given the species-common psychological vulnerability and the powerful stressors to which these communities are exposed, the most logical explanation for the felicidal actions of P02 and other cougars that first comes to mind, at least to a neuropsychologist, is complex post-traumatic stress disorder (c-PTSD), a formal diagnosis made for young African elephants who suffer a series of traumas—loss of mother and family, premature weaning, translocation, and absence of normative socialization by elders.[51] Male and female elephants exhibit symptoms consistent with PTSD, including inter- and intraspecific hyperaggression and killing, and infanticide.[52] The parallels with pumas are striking.

Documented cases of puma intraspecific killing that include infanticide are rare, but nonetheless the belief in cougar-on-cougar murder is prevalent.

A 2007 article admits that while "reports of strife, killing, and predation among mountain lions (*Puma concolor*) are not uncommon," significantly, "there have been no published reports of actual field observations of behavior during an event of intraspecific strife."[53] Notably, reported incidents are correlated with areas of intense hunting or highly developed, densely populated human-dominated areas, as in the instance of P01.

The most parsimonious interpretation of intraspecific killing argues that exogenous factors such as dramatic environmental changes (habitat destruction and loss, widespread and persistent killing) be taken into account. What is known with certainty is that chimpanzees, pumas, African lions, and all other wildlife whose behavior is cited as explaining the origins of lethal human conflict are also susceptible to psychological trauma.[54] Most studies date from after the 1920s, by which time wildlife habitats and populations had already sustained profound change from hunting and habitat appropriation and degradation. Yet no one has considered these extensive, historical neuropsychological effects of long-term disruption and violence on wildlife sociality and minds. Neither evolutionary biology (including its offshoot, evolutionary psychology) nor ethology has taken into account animal psychological vulnerability as a relevant factor in explaining intraspecific killing. It was only in 2005, when free-living African elephants were diagnosed with PTSD, that the vast accumulation of data documenting psychophysiological effects of trauma in nonhuman animals was openly brought to bear on wildlife biology.[55]

Viewed from science's expanded, interdisciplinary vantage, the puma family tragedy metastasizes into the horror of mass suffering. Taken together, the stories—of P22's anomalous incursions and those of other young males into human-dominated areas, of the pumacidal P02 who killed several conspecifics including his mate and bred with his offspring, of F51's terrified retreat into the cliffs after losing yet another litter and premature separation from her immature kitten, Lucky, and her hyperaggressive attack on M85—weave a disturbing narrative that counters the prevailing theory of evolution in control. Instead, the data add up to a series of tragic chapters in the ever-shrinking book of puma society.

Similar to the orca Tilikum, whose captive nightmare pushed him to betray essential orca values and kill three humans, the abnormal behaviors of these pumas reflect the breakdown of an ancient society and its moral decay as a result of human violence. If, as Darcia Narvaez writes, "moral conceptions and development emerge from the very conditions of . . . life, then it matters what those conditions are; habits and intuitions are built from experience."[56] By fundamentally changing the fabric of puma life, humanity has altered the moral principles of puma society. Our species has created a situation for pumas where "well-rehearsed stress states become traits. In this way neurobiological systems influence morality, setting up propensities to use different social and moral mindsets. What the brain's capacities look like has much to do with early life experience when brain system connections are being established."[57]

Without extensive, hunter-free habitats, carnivores have no place to run and nowhere to hide. At this late moment, it is unlikely that pumas will be able to regain their former crowning equanimity with their habitat and kin. Human excess has penetrated the minds of the great American lion and turned the puma paw of restraint into a weapon against itself. Like high-powered bullets, such impacts don't end at their immediate targets. Because stress transmits neurobiologically and socially, violence experienced by adults rips across time to strike at successive generations. Violence begets violence.

And what happened to F51's two seven-month-old kittens who were left orphaned when their mother died? In the following days, they struggled to survive. They were too young to have mastered the hunt. Growing weaker by the day, F75 eventually stopped, lay down, and lost consciousness. For a while, her sister, F99, stood over her, protective of the listless form. But she would not rise again.

It wasn't long before F99 joined her sister. Biologists found the young puma near a frozen carcass of an elk. Porcupine quills had penetrated her body and collapsed her lung. She had caught the prickly prey, but lacked the practiced art of avoiding quills. She was sixteen months old. Tracking data provide some glimpse into her final days. But perhaps the most poignant image is a trail camera video that caught her unawares. She is waking from an afternoon nap and lying next to the same elk carcass where she later will be found dead. She looks up, somewhat dazed. Her face wears the open naiveté of a kitten, an unguarded look that lasts far too long than is wise for someone her age and on her own. Does her countenance reflect the disorienting transition from sweet dreams to stark reality? Or is her sleepy insouciance an expression of a mind stunned by loss? Perhaps both.

7

Coyotes: The Predator Complex

Case No. 587
Name: Amber C.
Gender: Female
DOB: Unknown
Occupation: None
Marital Status: Single

Presenting Problems Extreme agitation from intrusive memories, dissociative reactions; psychological distress in the presence of flashing lights; fear of and hyperaggression toward caregivers; hy-

pervigilance; severe anxiety; non-normative eating habits; past self-injurious behavior; screams in the night; easily overstimulated.

History and Current Living Situation Amber was orphaned as an infant and has lived in foster homes or residential care facilities for most of her life. After reports documenting abuse while in foster care, she was relinquished and brought to a second foster facility. She appeared to have adjusted well there, and was socially high-functioning as well as in good health overall. Four years later, however, she sustained a gunshot wound to the head from an unknown assailant, as yet unidentified. Although her left eye was badly damaged and had to be surgically removed, the patient appears to have recovered well physically. But she has developed significant behavioral changes, including aggression and withdrawal into her room. She becomes agitated in the presence of flashing lights, apparently related to the light used by her assailant. Because of the risk that she will behave aggressively toward most caregivers and other residents, and may even cause harm, she now resides alone in her own quarters.

Mental Health Care History After several weeks, the patient was able to work with one therapist safely and on a consistent basis. Initial therapeutic goals included working with the patient in a manner that would not trigger physical assault or anxiety.

Mental Status Overall health condition has been maintained but shows strain. Does not engage in self-care. Nervous, suspicious, shows ambivalence to therapist. For example, expresses willingness and desire to be in close contact, but when therapist responds positively, subject suddenly becomes aggressive. No medication has been assigned. *Suicidal/homicidal ideation/plans:* Episodes of aggression and physical attacks on staff have attenuated, but the risk remains significant.

(Overleaf) Coyote spirit. Photo credit: Jami Hammer.

Alcohol and Drug Use (Past and Present) None

Domestic Violence Assessment (Past and Present) None

Diagnostic Impression (DSM-5) The patient presents with classic symptoms of Post-Traumatic Stress Disorder (PTSD), with no evidence of depersonalization or derealization. The head injury associated with the gunshot wound, however, raises the possibility of Major or Mild Neurocognitive Disorder Due to Traumatic Brain Injury (TBI) with Behavioral Disturbance. Symptoms of PTSD and TBI overlap, but in this case a differential diagnosis based on avoidance behavior (i.e., the flashlight) confirms PTSD. This does not preclude the role of TBI in the etiology of the patient's PTSD, as researchers increasingly recognize that PTSD can develop following both mild and severe cases of TBI, and that even mild cases of TBI may increase the risk of developing PTSD.

Diagnosis
309.81 Post Traumatic Stress Disorder
331.83 Mild Neurocognitive Disorder Due to Traumatic Brain Injury, with Behavioral Disturbance
Personality factors: Not assessed
Medical factors: Patient's vision is restricted to one eye
Psychosocial factors: V61.8 (Upbringing Away from Parents); V60.3 (Problem Related to Living Alone); V62.4 (Acculturation Difficulty); V62.89 (Victim of Crime)

Treatment Goals Socialization as a goal of treatment will, at the outset, reduce the risk of physical injury to Amber and her caregivers. Ultimately, treatment goals also include environmental modifications designed to reduce external stressors and triggers and foster enhanced subjective wellbeing. Specifically, treatment seeks outcomes that include reduced symptoms associated with PTSD, for example, flashbacks, anxiety, avoidance behaviors, and aggression.

Case number 587 refers to a young coyote (*Canis latrans*) named Amber, rescued by the Indiana Coyote Rescue Center (ICRC).[1] Her intake assessment is no anthropomorphic sleight of hand. It has precedence. Since the diagnosis of elephant PTSD, species-common assessments have been used to explore and describe great apes, parrots, and domesticated animals.[2] As in the cases of the bears, pumas, crocodiles, and the other species whom we have encountered, psychological descriptions reflect what neuroscience predicts: shared, cross-species brain structures and processes create mental and emotional states and expressions that will be familiar to humans. For Amber, this includes symptoms relating to trauma that accordingly may be assessed using standard methods used for humans.

Psychological symptoms must meet four criteria to formally qualify as non-normative (that is, not species typical): the symptoms and conditions need to be relatively persistent and expressed exclusive of any given specific context; cause an interruption or significant change in an individual's life arc; constitute identifiable psychological and somatic distress; or constitute significant behavioral alterations relative to an understood social and cultural space.[3] Amber's condition conforms to each of these criteria. Her post-trauma symptoms persist, the shooting has significantly changed what should have been a good life, psychological symptoms of hyperaggression and hypervigilance are pronounced, and her present behavior conforms neither to a healthful baseline relative to the "social and cultural space" of sanctuary captivity nor to that of free-living coyote society.

Amber's life at ICRC began in 2001. A veterinarian called authorities after he discovered that the so-called dog who had been brought in for examination was in fact a coyote. Amber was then sent to the Indiana rescue center that was founded by Cecilia Lambert. Similar to many who have established organizations to aid animals in need, Lambert had never thought of running a rescue center, let alone one for coyotes, until one day, well before Amber was born.

Lambert had been volunteering at a wolf education center when an attending veterinarian asked if she would be willing to raise two coyote puppies. She agreed and, not too long thereafter, in 1989, created a full-fledged rescue center.

Amber showing her eye injury after being shot in sanctuary.
Photo credit: Jami Hammer.

Lambert ran the ICRC until 2011, when she took ill and died a year later. After a brief interim, Jami Hammer took over the directorship and continues in that position today. It was Amber who led Jami to join the ICRC:

> I first came to the ICRC in October, 2005, while interning at Wolf Park about 40 minutes west of here. It all started with Amber. It was due to her shooting that I found this place. Ceann [Cecilia] had put up reward posters for information about Amber's shooting. Some interns and I saw it and wanted to help. After coming here—it was about nine months after Amber had been shot—I was hooked. I came back the following summer as an intern and lived here for a few years, then moved back home to Illinois to finish up some schooling, and moved back here permanently in October 2012, the year after Ceann passed away. During my time in Illinois, I continued to make weekend trips, nearly every weekend, to assist Ceann in her work, until the end. It's the best thing that ever happened to me and I will always thank Amber and Ceann for that.[4]

The wildlife rescue world is a Band-Aid, betwixt-and-between world. Rescue centers are first-response M.A.S.H. units that tend to collision victims at the intersection of wilderness and humanity. Eagles damaged from telephone lines, skunks hit by a car, wild turkeys peppered with buckshot—they are all frontline casualties who end up at the door of centers like ICRC.

Injured wildlife mortality numbers are high. An eagle with a broken wing cannot hope to live on her own; the skunk, strong of will but fragile of body, cannot usually survive the crushing blow of a passing car; and wild turkeys are often too crazed to be calmed by hands proffered to help and die from shock. Some rescued wildlife do get a second chance. Under the care of rehabilitative ministrations, they manage to survive and are released back into the wild. But many others are not so lucky. Unable to thrive on their own, a great number of survivors are either euthanized or spend their days in the confines of a cage.

Not all wildlife who come to rescue are wounded. In instances such as Amber's, a number of young animals had been taken in as "pets." Many find the idea of possessing a bit of the wild attractive. There is a burgeoning trade in native and exotic wildlife—lions, tigers, snakes, parrots, snakes, and other species—that is responsible for hastening the demise of already precariously endangered species.[5] These businesses have given an extra boost to an already vigorous enterprise. Exotic wildlife are permitted for private use, roadside zoos, and entertainment, and all sorts of non-native wildlife—cobras, crocodiles, piranhas, and you name it—manage to be sequestered legally and illegally in homes around the United States, with devastating effects on animal wellbeing.[6]

Native wildlife are another matter. In contrast to exotics, private possession of native wildlife is prohibited in all fifty states. Even temporary possession for purposes of rehabilitation requires a permit, license, or other authorization.[7] If a leg or wing amputation is required, regulations demand euthanasia in most situations. Illegally garnered wildlife usually end up being confiscated or abandoned. Wildlife born in the wild, but raised in captivity, are considered unsuitable for release, and law requires euthanasia unless they can be taken in by a valid wildlife sanctuary.[8]

For most of those who possess exotic or native wildlife, the honeymoon period of thrills rapidly fades as the animal begins to mature and act like the species they are. In the case of a coyote, this means howling, yapping, running around, chewing things, and at times hunting down the neighbor's cat or dog. As Hammer relates, "You get people who come across a coyote puppy when they are out hunting and think it would be fun to bring him or her home. But it rarely works out."[9]

Although the number of rescue centers has increased significantly over the past decades, there are still far too few to accommodate the demand.[10] Wildlife sanctuaries are jammed with the psychologically and physically wounded. By definition, captured animals are traumatized from the beginning. Most frequently, an infant is orphaned and captured when the mother or other family members are killed. Capture is shocking in all instances, transport is lethal for most, and, together, including the arrival to the unnatural destination, comprise a multi-trauma assault. The fate of individuals who are born into captivity is not much different.

As neuroepigenetics points out, when you fool with Mother Nature or mother's nature, the results are usually disastrous. Almost to a one, captive and captured wildlife are beset with a suite of psychophysiological ailments, including tendencies toward self-injury and infanticide, depression, hyperaggression, deformities, compromised immune systems, and disease.[11] Unless appropriate care and environments are provided, native wildlife usually fare poorly in the liminal space of human possession. Even under the optimized oversight of rescue centers and sanctuaries, while residents may appear to live good, fulfilling lives, it is, as Elephant Aid International founder Carol Buckley reminds us, "still captivity."[12]

When denied agency—the ability and opportunity to make important life decisions—all animals suffer. The severe and lasting symptoms associated with social, physical, and psychological restraint—whether the restrained is a human prisoner or a nonhuman animal exhibited in a zoo, aquarium, or circus—are pernicious.[13] In addition to juggling the needs of the flood of animals who require help, the greatest challenge facing wildlife rescue centers is mitigating the negative effects of captivity. Jami Hammer talks about the difficulties of providing coyotes, who live at the sanctuary and cannot be released, with *environmental enrichment,* that is, activities that will engage an individual cognitively, emotionally, and physically.

Interestingly enough, environmental enrichment is an active topic in both neuroscience and animal research. It has long been recognized that the brain is most plastic in youth and that neuropsychological patterns are shaped, as we saw illustrated in detail with grizzly bears, through epigenetic alterations that affect cognitive and affective outcomes. The tenacity of early experiences makes any changes in brain and mind difficult as one ages, but of particular concern to neurogeneticists is the question of how, or if, neural damage caused by Parkinson's, Alzheimer diseases, or injury might be healed.

Using the ubiquitous lab rat as a human surrogate, scientists have discovered that effects from traumatic assaults to the brain, whether induced chemically or by maternal separation, can be compensated to a certain extent if the rat's environment is enriched appropriately. In the study, infant rats were separated from their mothers. Premature maternal separation increases stress reactivity, that is, decreases an individual's ability to withstand stress such as food deprivation. But the effects of maternal separation in those who were later housed in environmentally enriched cages (they lived with eight others in a series of large 60 × 30 × 60 cm cages that were interconnected with a burrow system and "filled with toys that were replaced regularly") were reversed, while those exposed to "standard laboratory conditions" (they lived with a maximum of two individuals in 20 × 40 × 30 cm clear plastic cages) showed no such compensation or reversal.[14] Other studies found that environmental enrichment positively affects neurogenesis and reduces "spontaneous apoptotic cell death" (what is also called programmed cell death) in the rat hippocampus by 45 percent.[15] In a less cruel and more entertaining study, researchers found that human brains also profit from environmental enrichment.

It turns out that driving well and efficiently in Britain's capital requires detailed knowledge of all the ins and outs of streets, finding the quickest path given changes in traffic conditions and so forth—what local drivers refer to as "being on The Knowledge." Acquiring this know-how takes two years of training and then passing rigorous exams in order to qualify for licensure as a London cabbie. When researchers compared scanned brain images of these seasoned cab drivers with those of the enriched rats, they found out that they have something in common that ordinary non-driving humans don't: enlarged hippocampi. The posterior hippocampus is a part of the brain that is associated with spatial memory and the ability to navigate well, whether it be in the air, water, or land. So rat or human, lab maze or London metro, keeps the brain in tip-top performance shape.[16]

Zoos, aquaria, personnel, and conservation biologists are also concerned with enrichment. Since animals became a research tool, psychologists and zoos have appreciated the devastating effects of captivity. Psychologist Harry Harlow, famous for his self-coined terms "rape rack" (an apparatus for artificial insemination) and "pit of despair" (a chamber where infant animals were kept in isolation for months on end), used this understanding to illustrate the effects of environmental deprivation on laboratory nonhuman primates.[17] Heini Hediger added to the discussion with his classic 1955 book *Studies of the Psychology and Behavior of Captive Animals in Zoos and Circuses,* which focuses on the emotional and mental symptoms caused by capture and captivity of tigers, lions, great apes, and other wildlife used for human entertainment.[18]

Now that captive breeding has become a cause célèbre, where, on one side, proponents insist that it solves the global species extinction crisis and, on the other side, critics argue that such programs are money-making schemes for zoos whose funds would be best spent on badly needed efforts to conserve in situ, captive health has become a central topic. Despite epigenetic advances in the laboratory, however, the fact remains that all captive-held wildlife, who by definition are denied normative social interactions and the expanse of natural habitats, live lives of profound deprivation. No interconnecting burrows, toys, or patched-together assemblage of conspecifics can make up for what they have lost or never had—freedom, agency, and the chance to live a life on their own terms with their loved friends and family. Jami Hammer sees this every day in her efforts to make up the emotional, cognitive, and physical shortfalls for the coyotes who reside at the Center.

> I've worked with coyotes for twelve years now. Believe me, it is by far easier to interact with wolves and domesticated dogs. Coy-

> otes are amazing. . . . Coyotes are just something else. They are extremely energetic and their personalities are kind of a mixture between cat and dog—very intense and very playful. Coyotes are always on the go and very sensitive to overstimulation. For example, when you pet them, head or back, you can't keep on doing it like you might with a dog, most of whom find the rhythm of stroking soothing and relaxing. A coyote is like a perpetual puppy in that way—they get more and more excited because their senses are always "on." They get bored easily. We've designed all sorts of things to keep them engaged and fulfilled. In fact, that is the hardest part of the day for me—thinking up new ways to keep the coyotes happy and "doing their jobs." You think of what a coyote would do free—running all over, being with family, finding food, smelling and seeing all sorts of wonderful things—then compare that to living in an enclosure. We come up with some ways that really seem to improve their psychological wellbeing. For instance, they love the "obstacle course" we set up, they jump up on something and love to balance, and just go berserk. Coyotes are also really "mouthy," meaning they prefer chewing/mouthing on your hand to a toy. Unlike other rescue wildlife, there's not that much known about coyotes because people hate them. They get shot, poisoned, or killed in some other way and people don't even think of rescue or rehabilitation.[19]

Along with rattlesnakes, skunks, raccoons, possums, and other so-called nuisance wildlife, coyotes tend to slip through the cracks of fame, fortune, and sympathy with unhappy results. When Europeans arrived in the New World, coyotes immediately failed to pass the test qualifying them as "good" animals (like bison, elk, and antelope who could be hunted and eaten) and were immediately relegated to the category of "bad" animals (those inedible animals who were audacious enough to eat the same prey as humans).[20] One classic book on another predator, cougars, explains: "The Christian obsession with morality greatly influenced the European's view of native cultures and wilderness . . . They saw it as their moral duty to civilize not only the 'savages' but the land itself. There was no place for predators in such a world."[21] As a species assigned to the predator class, coyotes definitely strike a negative chord among modern humans and hence rate high on the least-wanted list.

Statistics on coyote and other wildlife killing are difficult to obtain.[22] In a 2012 newspaper exposé, one reporter investigating the activities of the U.S. Fish and Wildlife Service quipped that "while even the military allows the media into the field, Wildlife Services does not." The reporter provides examples of how inquiries into wildlife "control" have provoked hostile retorts by wildlife agencies. In response to the reporter's request for a ride-along field observation with agency personnel, the director of Nevada's state wildlife service, Mark Jensen, replied, "If we accommodated your request, we would have to accommodate all requests." Acting California Wildlife Services state director Dennis Orthmeyer said about his organization: "We pride ourselves on our ability to go in and get the job done quietly without many people knowing about it." And an Elk Grove, California, Wildlife Services manager, commenting on the number of beavers the agency had killed, exclaimed, "This information is Not intended for indiscriminate distribution!!!"[23]

Coyotes are killed routinely by the thousands every year.[24] In Pennsylvania, where an estimated 100,000 coyotes live, annual hunts kill 40,000.[25] Unlike domesticated animals, wildlife in the United States are not governed by any welfare laws. An incident in Michigan in 2015 where a hunter shot a coyote three to four times, then let his dogs loose on the dying animal, displayed appalling cruelty but was perfectly legal.[26] The details of Amber's story illustrate the distaste that coyotes seem to evoke in many.

After her arrival at the Indiana sanctuary, Amber quickly matured under the loving care of Cecilia Lambert. The young coyote was playful, relaxed around humans—even at first meeting—and was all-around well-adjusted to the semi-domestication scene. But four years into her tenure, everything changed: "It happened in 2005. What we can piece together is that someone trespassed and came onto the grounds in the middle of the night. The coyotes would have barked but they do that often—they are active at night—so we did not think anything was wrong. The shooter must have had a flash-

Indiana Coyote Rescue Center director Jami Hammer with rescued Coyote. Photo credit: Jami Hammer.

light, or one attached to the gun scope that he shone in her eyes. At that time, Amber loved everyone so when she got wind of him, she probably ran right up to the front of the enclosure to greet the stranger. She was always so friendly, we called her the Ambassador. But, this time, she was met with a gun. She was "shot in the face with a rifle at point blank range."[27] Severely wounded and bleeding profusely, Amber was rushed to the emergency hospital for treatment. Though she survived extensive tissue damage, the surgeon had to remove her left eye. As the case notes indicate, she has largely healed physically, but the trauma and shocking circumstances of an assault in her own home changed Amber irrevocably.

Today, Amber has vastly improved relative to the first two years after the shooting. With focused and careful work, Jami was able to ease Amber's self-biting and obsessive overgrooming. While this represents a huge step forward, Amber remains aggressive and exhibits insecure ambivalence—at first she welcomes someone's approach, then lunges and growls. If she catches the glare of a flashing light, she becomes terrified. Most of the time, Amber appears ill at ease and frightened in the presence of strangers. Even years later, Amber will not permit anyone to enter her enclosure with the exception of Jami and, in a twist of fateful symmetry, her beloved Oswald "Ozzie" Bracero.

Bracero and Amber have much in common. He may be a human hailing from the steel canyons of Brooklyn and she, a coyote from the woods of Indiana, but both suffer from PTSD and both have had their left eyes injured. Bracero was a first responder when the New York City Twin Towers fell in 2001, part of the desperate attempts to dig bodies from the chaotic debris and crushing smoke. Later, in 2006, on tour in Iraq, his vehicle was blown up by a roadside bomb and his convoy was attacked. The explosion caused irreparable damage to his left eye from bomb shrapnel. After briefly recounting Ozzie's history, Jami adds, "It was that coincidence, that common recognition that they had both sustained damage to their left eyes, which is part of the reason he took such a liking to her. Normally, Amber is afraid of men, but with Ozzie—she's completely different. She has never shied away from Ozzie." For city dwellers, the lonesome howls and barks of the coyote may not be music of the spheres to sleep by. But not so for Bracero: "Ozzie says he has never slept as well as he has here since he started struggling with his PTSD. Even amidst the howling and fox screeches through the night, he sleeps and sleeps well. The coyotes here tend to have a healing effect."[28]

Second only to eliminating the source of trauma or stress, the most important step in recovery is cultivating the kind of trusting bond that Bracero has developed with Amber. This is particularly crucial for a captive, because, by definition, the caregiver is the world, the provider of all and key to life itself. The captive depends on the caregiver for food, water, social contact—everything. As such, this bond is implicitly volatile, especially for those whose human-caused wounds have sent them to captivity. The human plays ambivalent roles—captor and caregiver—with the power to turn life into death at any moment. Subsequently, every move, every gesture speaks volumes to the captive. Building a lasting trust and a sense of safety in confinement takes patience and time, enough time so that the captive will develop confidence in this new reality. When this occurs, the energy directed to defense may be redirected to healing.

The majority of coyotes who come to the Indiana Center bear the wounds of psychic pain:

> They develop stereotypic pacing, head rolling, they are depressed and sleep excessively. Some will jump up on the fence, stretching and trying to look for a way out. One of our residents used to bite at his tail and chewed all the fur off it. Most of the time, they are not aggressive, but we work hard so that symptoms of distress abate or do not develop and we are successful with most of the residents. Even without trauma like Amber's, coyotes can develop stereotypic behaviors from the stress of confinement and insufficient enrichment and attention that mediates the effects of captivity. So we spend a lot of time with them, work hard to make their days and enclosures rich and interesting, and most share an enclosure with another coyote . . . three share one and can become really close and very good friends. This helps a lot. People don't think that coyotes have emotions or are socially sensitive, but they are. You can see it in the wild, the way they are with each other and their families. Usually wolves are showcased as model parents, but I think that coyotes are even more amazing.[29]

Although attitudes toward coyotes are largely informed by such characterizations as the cartoon figure Wile E. Coyote (who was voted as one of the top sixty nastiest villains by *Reader's Digest*), in real life, coyotes conform more closely to Disney family-values criteria: mothers who would die rather than abandon their young, fathers who faithfully pitch in with a thirty-something parenting style, and, always a crowd pleaser, parents who mate for life.[30] Similar to bears and pumas, coyotes are born into and develop in the warmth of a secure attachment, surrounded and supported as they are by parents and siblings in the natal den.[31]

If coyotes are so warm and fuzzy, why does their bad image persist? Statistics show that coyotes, like other carnivores, seldom injure humans, make barely a depredation mark, and basically are all-round good guys who help keep the ecological balance in their environments. Predator control advocates, however, insist that without routine carnivore culls, deer and other game species numbers will drop precipitously.

Agency insiders disagree, and claim that this notion has perpetuated as a myth within a myth. According to John Shivik, formerly the leader of the USDA/APHIS/WS National Wildlife Research Center Predator Research Station, and now U.S. Department of Agriculture Wildlife Program Leader, Intermountain Region, the misconception that carnivores are responsible for low deer numbers dates from the era when deer hunting was unregulated. Consequently, like carnivores themselves, deer and bison were almost entirely eliminated from the western United States as a result of unlimited hunting. Only when legal restrictions were implemented concerning when, where, who, and how game could be hunted was the hemorrhaging staunched and herbivore populations able to recover and rebound. Shivik draws on his familiar territory, Utah, as an illustration.

By the 1950s, after hunting regulations were imposed, Utah's deer population increased to a robust 500,000. A few decades later, numbers plummeted to half that. The cause was not nonhuman predators but the left-right punch of bitterly cold winters and habitat appropriation by human development. Increasingly, deer have had to cope with the progressive loss of winter habitat that has been paved over at a rapid rate to keep up with the demands of a skyrocketing human population. Today deer numbers hold at 300,000, but the outmoded half-million figure remains a fixed standard, and the effects of human encroachment are ignored.

Predator control advocates also maintain that killing wild carnivores is not only necessary but is also by far the most effective method of keeping nature balanced—again, not so. An analysis by the Natural Resources Defense Council states that the U.S. Department of Agriculture's "own statistics show that most livestock losses come from weather, disease, illness, and birthing problems, not predation," yet over $100 million is spent annually to kill over one million native carnivores "even when the effectiveness of such killing is unproven or, worse, counterproductive.[32] And an Idaho study described as the "most extensive large-scale assessment of predator control effectiveness" concludes that climate—not coyotes—dictate deer population numbers: "winter severity in the current and previous winter was the most important influence on mule deer population growth."[33] Now grounded in an ecosystem perspective, the scientific consensus on predator control as a means to enhance deer populations is that "killing predators does not always and automatically create more deer."[34] Furthermore, tinkering with carnivores upsets the workings of a finely tuned evolutionary clock that has far-reaching consequences for the entire ecosystem.[35]

Coyotes meanwhile, unaware of the debate, seem to have been able to outsmart predator control efforts. Historically coyotes' range extended from southern Canada to northern Mexico, and the western Great Plains and Midwestern United States. But in contrast to most other carnivores, the species has expanded its territory in nearly all directions—as far east as Long Island, New York, as far south as Panama, and as far north as Alaska.[36] In addition, they have increased the number and types of ecological niches that they are able to use. Formerly found largely in grasslands and deserts, coyotes now seem to thrive in highly urbanized areas and agricultural lands.[37] Although sightings of coyotes were once regarded as "freak" events in New York City, they have now been spotted in every borough except Brooklyn.[38] There is even evidence of coyote-wolf hybridization, which illustrates the species' social and genetic, as well as ecological, flexibility.[39]

But if the effectiveness of large-scale predator control is dubious, why is so much money spent on it? Why, if carnivores are not the problem they are made out to be, and their eradication does not make ecological, ethical, or economic sense, does the practice persist and with government backing to boot? Standing wildlife policy and rhetoric violently clash with both trends in public opinion and science's consistent and inarguable statistics and studies that reveal the fallacy of the predator myth. Whereas in the eyes of newly arrived Europeans "certain animals such as the wolf, rattlesnake or coyote were vermin, murderers, gangsters, and their elimination was perceived as progress and the beneficent altruism of one who considers his community as highly as his own welfare," the majority of today's urban and suburbanized humans are more apt to pick up a camera rather than a gun when meeting Wile E. Coyote or Red Riding Hood's nemesis. Statistics attesting to this sea change in public attitude bear this out.

The impact of humans on predators is huge in terms of the numbers of wild animals killed annually and the number of near species extinctions in North America's past and present, but the number of people responsible is disproportionately small. According to the U.S. Fish and Wildlife Service's *National Survey of Fishing, Hunting, and Wildlife-Associated Recreation* published in 2011 and updated in 2014, just 13,683,000 Americans age sixteen and older hunted, which, given the total U.S. population in the same year, 311,591,917, amounts to a mere 4.9 percent of the American population (or 5.7 percent of the American population age sixteen and older).[40] Fifty-seven percent of all hunters hail from rural areas, 89 percent are male and 94 percent are white, only 3 percent earn less than $20,000 and 8 percent earn between $50,000 and $100,000 annually. In contrast, during 2011, about 71,876,000 Americans, or nearly 23 percent of the total U.S. population, engaged in (non-lethal) wildlife watching.[41] These numbers show that inordinately few humans kill an inordinately many wild animals. It is also clear that despite the recognized shift in public attitudes, "the predator, particularly the wolf and coyote, continue to bear this historical stigma."[42] The mass killing continues. Finding out why requires a bit more probing and a first clue begins by following "le buck"—not the antlered kind, the green kind.

When the pioneer killing culture became law at the turn of the twentieth century, killing carnivores became big business. Predator control agents confessed that they purposively refrained from killing a long-term "breeding pair" of coyotes because "it would be like killing the goose that laid the golden eggs. You always want to leave a breeding pair out there to raise litters and keep the sheepmen thinking coyotes will drive them into bankruptcy if predator control is ever halted."[43] Paul Maxwell of Grand Junction, Colorado, former government trapper turned coyote advocate, admits that this realization forced him to quit his day job: "This killing business is out of hand . . . I grew up on a cow ranch, and as a kid I remembered my uncle would no more kill a coyote . . . Now the ranchers, the Fish and Game people, and everyone has just plumb gone insane. All at once they're saying the coyotes are killing off the deer and the antelope. For thousands of years these animals lived together. Now suddenly the coyotes are killing them off? They're just figuring every angle they can think of to keep a bunch of bums on the payroll."[44] In accordance with Utah's Mule Deer Protection Act, "There is also good money to be made from bounty hunting with incentives to kill provided by state and federal agencies." The state "currently obligates $500,000 per year to be paid to people who submit coyotes for a $50 per coyote reward. Wildlife Services receives an extra $600,000 to gun coyotes from aircraft, an effort funded by a $5 fee attached to big-game licenses."[45]

Benefits from killing are also accrued by the agencies that oversee wildlife. John Shivik, also author of *The Predator Paradox,* has witnessed this firsthand as a government employee. "Modern wildlife management was founded by hunting advocates in a more simple time, but the economic drivers have been retained. How we go about solving a problem is influenced by who pays for it. The historic funding source for most state wildlife agencies is consumptive users via hunting license sales. It shouldn't go unnoticed that one-third of state agencies in the United States use the word 'game' in their title rather than 'wildlife.'"[46] Hunting is an "economic force" in the United States. The U.S. Fish and Wildlife Service reports an average yearly $1.4 billion in revenues from hunting and fishing licenses and tags. This does not take into account the $11.8 billion in taxes generated (in 2011) by motels, restaurants, and weapon and clothing sales.[47] Profits are comparably substantial in private sectors. The "annual federal income-tax money generated by hunters' spending [in 2002] could cover the annual paychecks of 100,000 troops. That's 8 divisions, 143 battalions, 3,300 platoons and some major money."[48] In 2005, Leonard Ellis sold his hunting business in Bella Coola, British Columbia for a tidy $1.3 million.

The new trend to breed trophy carnivores in captivity has added its fuel to the fire. A 2007 study by the Texas AgriLife Extension Service, which is linked to Texas A&M University, estimated the exotic game industry in the state of Texas alone "to be worth $1.3 billion."[49] Although many argue that trophy and canned hunting (a trophy hunt where a lion, tiger, wolf, or other wildlife are confined for easy killing) save species, the killing of up-to-$350,000-a-head rhinos, lions, and other wildlife has driven the species even closer to extinction and does not appreciably support local communities.[50]

The carnivore myth was able to achieve near immortality only by the tight link between and co-evolution of the private hunting sector with government wildlife agencies. Mandate (carnivore eradication) and method (government-sanctioned and subsidized killing) became fused by institutionalizing who has authority over the nation's wildlife and how they should implement that authority. On the surface, carnivore protection appears to be an active concern of government authorities. Two years after the Endangered Species Act was passed, then-U.S. assistant secretary of state Nathaniel Reed proclaimed, "We are dedicated to the belief that America has matured to the point that we are no longer willing to sacrifice the end product of eons of evolution—a species or subspecies of wildlife—on the altar of the god called Progress without putting up one darned good fight!"[51] The history of U.S. wildlife policymaking and its architects, however, reveals the chameleon deftness that has created an institutional barrier impermeable to changing ecological and political currents that go against carnivore cleansing.

Over its nominal progression from its establishment in 1885 as the U.S. Department of Agriculture's Branch of Economic Ornithology and Mammalogy, whose purpose was to deal with bird "damage" to agricultural crops, to the Division of Ornithology and Mammalogy, then onto the Bureau of Biological Survey in 1905, the agency's "mission crept accordingly, with rodent control added to the list of responsibilities in 1913."[52] Two years later, predator control began and in 1924 the organization was given the new title Division of Predatory Animal and Rodent Control.

The Animal Damage Control Act of 1931 proved to be a fatal turning point, the moment when "the transition from mission creep to federal takeover of state management of predators" occurred.[53] The act meant that federal funds and authority would be given for the mass "destruction of mountain lions, wolves, coyotes, bobcats, prairie dogs, gophers, ground squirrels, jackrabbits, and other animals injurious to agriculture, horticulture, husbandry, game, or domestic animals, or that carried a disease."[54] After more "name changes [and] bureaucratic jockeying," which included yet another title, Animal Damage Control (ADC, nicknamed by its detractors "All Dead Critters"), the agency took its present ironic moniker, the Wildlife Services program of the USDA Animal and Plant Health Inspection Services (APHIS). Its mission is "to provide Federal leadership and expertise to resolve wildlife conflicts to allow people and wildlife to coexist." Tellingly, the website landing page with mission statement leads with the large, bold-fonted heading, "Wildlife damage."[55]

The mixed message of the ADC and its predecessors is long-standing. At one time, the ADC was housed within the Fish and Wildlife Service, which administered the inversely envisioned Endangered Species Act. Irreconcilable differences were acknowledged, and by 1985 ADC was transferred back to the U.S. Department of Agriculture under the Animal and Plant Health Inspection Service (APHIS), an agency better known for veterinarians who inspect the meat supply, and for officials who enforce the Animal Welfare Act, which does not cover wildlife.[56]

The persecution-protection duality is illustrated by the whiplash listing and delisting of the gray wolf (*Canis lupis*). The gray wolf and the grizzly bear are two of the most labile of endangered species when it comes to a conservation status variously hovering between a "species of concern" (so named by the International Union for Conservation of Nature [IUCN]) and a hunting target. In 2011, the U.S. Fish and Wildlife Service delisted wolves in the Western Great Lakes Distinct Population Segment (DPS) because "wolf numbers and distribution in the Western Great Lakes DPS have exceeded the population criteria identified in the recovery plan." In making the change, the service assured the public that the current wolf population, estimated with amazing precision at 2,921, will enjoy "continued survival . . . due to protections afforded on lands that are federally managed by the National Park Service and the U.S. Forest Service and due to management provided under the Minnesota Wolf Management Plan."[57]

When, however, the U.S. government moved to delist the wolf, it came under fire for dubious maneuvers to paint the truth: "After the FWS [U.S. Fish and Wildlife Service] was criticized for meddling with its own peer review panel (i.e., removing scientists who had publicly spoken out against the proposal)" an independent panel of scientists was convened at the National Science Foundation's National Center for Ecological Analysis and Synthesis in Santa Barbara, California.[58] The panel determined that "the science used by the Fish and Wildlife Service concerning genetics and taxonomy of wolves was preliminary and currently not the best available science."[59]

Great Lakes wolf hunting and trapping proceeded until 2014, when a federal judge ruled that "the removal was 'arbitrary and capricious' and violated the federal Endangered Species Act." By then more than 1,500 wolves had been killed.[60] U.S. Fish and Wildlife Service spokesperson Gavin Shire responded that the agency "was disappointed" because "the Great Lakes states have clearly demonstrated their ability to effectively manage their wolf populations."[61] Nonetheless, shortly thereafter, a third bill was reintroduced to delist the wolf.[62]

Always regarded as the aloof, impartial third party, science can get caught up in the larger political context as well. Scientists committed to carnivore conservation walk an edgy line as they try to stay in the good graces of those agencies that grant permits to conduct research on public land. Furthermore, the roles of scientists on the inside and those on the outside are sometimes blurred by conflicting agendas. For example, after the 2011 delisting, over 20 percent of Wyoming's wolf population was shot in less than two-and-a-half months—and some of these wolves had been collared and tracked as part of an ongoing 1999 research study in Yellowstone and Grand Teton National Parks. According to U.S. Fish and Wildlife Service wildlife biologist Mike Jimenez, who is also the wolf management and science coordinator, "We always knew wolf-hunting would come, and the service has always encouraged it . . . It's part of having very successfully reintroduced wolves to this region. When they're outside the parks, they're on public lands—and the reality now is that when they're out there, they're going to be hunted."[63]

The ambiguous role that government agencies play has many wondering whether collared wolf kills are coincidences. In Oregon, although the Department of Fish and Wildlife (ODFW) maintains that they do "not give exact locations in real time," the state agency sends thirty-eight ranchers "daily text messages that tell them where collared wolves in the Imnaha pack were the night before." Rancher Ramona Phillips confirms that, in addition to posting specific information about each wolf on its website, ODFW passes on the location of radio-collared wolves: "ODFW texts us every morning and tells us where the collared wolves have been that evening. In one case they texted us and said we had a collared wolf coming onto our property."[64] Oregon has delisted the gray wolf.[65]

Clearly many scientific methods employed in the name of conservation are of questionable benefit to their subjects. Radio-collaring data that have been so useful in deconstructing carnivore myths carry with them a downside. In addition to providing the animal's location to possible hunters, animals experience both short- and long-term psychophysiological effects from darting and radio collaring, as researchers discovered when a still-dazed grizzly recovering from darting uncharacteristically mauled a Yellowstone hiker, Erwin Evert. In 2102, a "USGS crew had trapped, tranquilized, studied and released a 430-pound grizzly bear . . . [that] was likely antagonized by its experience with the researchers."[66]

Minnesota governor Mark Dayton banned all radio-collaring of moose due to these kinds of concerns. In one study, of the seventy-four infant moose who had been radio-collared right after birth, an alarming number died as a result of researchers' handling and subsequent mother abandonment. It is thought not uncommon for a deer or moose to abandon her young after human handling.[67] Six adult moose, too, died from the use of tranquilizers over a single winter.[68] The effects of stress and trauma cannot be underestimated. A video of an AeroTech employee hired by the state of Maine to aerial-track then dart and hogtie a moose to affix a radio collar shows the tremendous emotional impact on the infant ungulate as she, obviously still affected by the drug's side-effects, nonetheless tries to fight off her supposed predator.[69]

Many injuries and mortalities related to wildlife capture go unreported because of an "increasing public sensitivity to animal welfare."[70] Pumas, especially those with kittens, will fight with pursuing hounds who can be very aggressive and attack the big cat. If the hound pack is large, inexperienced lions and kittens may be killed. Injuries can also be sustained during life-or-death struggles to evade capture, and a dart-sedated lion may fall from the tree or boulder on which it was found.[71] Beyond the risk of physical injury, the chase can be energetically and psychologically traumatic. For all intents and purposes, pumas do not know if they are being chased for the kill or in the name of science (which is also not always a benign reason, as illustrated by the Nile crocodile massacre by conservationists). A short video taken by a professional mountain lion houndsman hired by researchers at the University of California, Santa Cruz, shows a cougar exhausted from the chase, lying down, head hanging and panting hard.[72] In 2010, mountain lion hunting was banned in the state of California. Two years later, the state banned the use of dogs to hunt bears. The legislation, however, was amended to allow dogs to be used for big cat research.[73] As mentioned earlier, drones, also known as unmanned aerial viewers (UAV), which are now increasingly used by biologists to follow wildlife, have been shown to cause extreme psychophysiological stress as indicated by elevated heart rate and depressed heart rate variability.[74] Similar to the children who endured the 1993 siege in Waco, Texas, and whose trauma was apparent only in anomalous heart rate variability, black bears may show few overt behavioral changes from UAVs but their heart rates indicate that they find encounters with drones stressful.[75]

The language surrounding human interactions with wildlife also obscures what is really happening. Orwellian doublespeak is pervasive. For example, the terms "conservation" and "management" are used interchangeably, as are their methods. So-called conservation tools, for instance, include hostile manipulation, government-approved hunts, pest and reprisal killings, sport harvesting, capture-translocation programs, and aversive behavioral conditioning. Furthermore, the brutal reality of mass killing is coded to make it more palatable: "Kill euphemisms are tailored for the style of hunt. Trophy hunters . . . favor 'taking,' or 'collecting,' a nod to the golden era of safari hunting, when celebrated British nobles dragged entire families of zebra and gazelle back to their gloomy castles as carcasses. . . . 'Harvest,' with its undertones of a bygone era of ripe wheat fields and feasting pilgrims, has become the rhetorical weapon of choice for hunting organizations liaising with the American public.[76] On its website, the Arizona Game and Fish Commission slips the word in with two saintlier aims: listing as its mission only the 'management,' 'preservation,' and 'harvest' of wildlife. Nowhere in the statement does the word 'killing,' or even 'hunting,' appear."[77]

The parallels with war are chilling. Lee Burkins, a Green Beret veteran of the Vietnam War, maintains it is linguistic convention that is the first to fall in war: "It is said that 'truth is the first causality of war' but veracity is wounded long before the battlefields are born. The first shot is fired when language is twisted and sometimes invented for the benefit of authorities to justify their actions, usurp laws meant to keep peace and convince others to fight."[78] Just as divergent etymologies of "peacekeeping" and "war" have merged into one meaning, so has killing and conservation.

Grizzly conservation plans provide another example. In 2013, "an interagency team of managers and scientists" drafted the Northern Continental Divide Ecosystem (NCDE) Grizzly Bear Conservation Strategy, whose purpose is to "describe the coordinated management and monitoring efforts necessary to maintain a recovered grizzly bear population in the NCDE . . . The [participating] agencies are committed to be responsive to the needs of the grizzly bear through adaptive management actions based on the results of detailed annual population and habitat monitoring."[79] The plan enumerates multiple laws that grant authority to government agencies for diverse practices, including a Montana law that "provides authority to the MFWP [Montana Fish, Wildlife and Parks] commission to set rules and regulations for grizzly bear hunting. The MFWP commission has the authority to: fix, open, close, lengthen, or shorten hunting seasons; declare hunting arms specifications; set possession and bag limits; set tagging and license requirements; set shooting hours; open special areas, and issue special licenses to manage grizzly bears through sport harvest."[80] Conserving wildlife apparently requires killing the animals, but only to the point of preserving enough bodies to kill in perpetuity. Originally coined to communicate the need for preservation and protection of nature, conservation has become a salvo in the war against wildlife.[81]

Public lands specifically reserved for wildlife offer little real protection. Established in 1903 by Teddy Roosevelt, the U.S. National Wildlife Refuge System that now encompasses over 150 million acres has as its mission to "administer a national network of lands and waters for the conservation, management, and where appropriate, restoration of the fish, wildlife, and plant resources and their habitats within the United States for the benefit of present and future generations of Americans."[82] But the nine principles articulated in the Wildlife Refuge System Improvement Act of 1997 include this statement: "Wildlife-dependent uses involving hunting, fishing, wildlife observation, photography, interpretation, and education, when compatible, are legitimate and appropriate uses of the Refuge System."[83]

Coyotes playing at the rescue center. Photo credit: Jami Hammer.

Proponents of the kill-to-conserve philosophy are intolerant of any challenge to the wildlife-human barrier. Good wildlife are fearful wildlife. A former law enforcement officer with a fish and game agency opines, "The image of an angry dangerous bear is essential to sell the public that killing is necessary. If people find out that Smokey [the Bear] is really a nice guy, then there goes hunting fees, there goes your job . . . Plus, most agency folks have a lot of personal and ego investment in being macho. It makes them look pretty silly if people realized that bears don't go out of their way to hurt you. That's why they go after individuals who don't pay attention to the myth and show the public that, yes, bears, pumas, and coyotes are nice! In the eyes of the agency, a human crosses that line and he or she is no different than the bear."[84] In other words, the law reinforces species apartheid.

Those who stray too close to the species line are considered to be fair game as much as the animals they befriend. Any breach of the well-guarded physical and psychological boundary that keeps wildlife at bay is countered with force. There are few issues more charged than the topic of consorting with another species, particularly carnivores. Gestures of interspecies détente are frowned upon as a bad habit that leads only to grief for both species. People are expected to side with their own kind, not with neighborhood wildlife. A few examples of what happens to bear-friendly people illustrate the point.

Eighty-one-year-old Mary Musselman, a retired gym teacher in Sebring, Florida, was sent to jail for feeding black bears. After her release, she was returned to prison because she broke her parole by putting out bread crumbs for crows at her home. The court ordered Musselman to live in an assisted living facility, where she died a few months later.[85] Elsewhere, after years of tug of war, Lynn Rogers, considered the foremost expert on black bears, had his research permit revoked by the Minnesota Department of Natural Resources for feeding bears despite decades of data documenting that providing snacks for bears does not create a frenzied ursine foodie.[86]

Farther north and west at Christina Lake, British Columbia, despite almost two decades of the practice with no harmful incident, Allen Piche was charged with violating "the Wildlife Act, which maintains fines for feeding dangerous wildlife of up to $100,000 for a first offence, and/or a jail term."[87] He was found guilty, and later, conservation officers informed the media that "17 bears had been shot in the southern Interior that spring, because they were too habituated to humans. Some were hanging around the popular Christina Lake campground areas scavenging with no fear of humans."[88]

The contradiction between science and fiction is perhaps nowhere starker than in the case of Alaskan Charlie Vandergaw. More than twenty years ago, the retired teacher bought land in the Matanuska-Susitna Valley that pours from the base of the glaring white-peaked Alaska Range to the cerulean blue waters of the Cook Inlet north of the Kenai Peninsula. The valley, nicknamed "Mat-Su," is a popular tourist destination. Its "towering mountains, huge glacier valleys, fish-filled rivers and lakes, abundant Alaska wildlife, glorious hiking country, scenic camping and quaint frontier communities," as well as its close proximity to easily accessed Anchorage, makes the valley the place "where Alaska comes to play."[89]

Vandergaw built his cabin in a remote corner of the valley. Over the years, and with no invitation, black and grizzly bears began showing up to what eventually came to be called Bear Haven. As the seasons rolled by, Charlie and the bears became friends and, doing what friends and family usually do, broke bread together. Their summer get-togethers became an annual ritual when Charlie came up after the spring thaw. That grizzly and black bear families lounged in proximity attests to the relaxed and secure atmosphere that Bear Haven offered. Under other circumstances, when times are lean and the forests become menacing hunter blinds, grizzlies may stalk a black bear for a meal.

Photos during these early halcyon years show Charlie's arm over the backs of one or two full-grown grizzlies enjoying the lazy hazy days of summer, and holding black bear cubs whose proud mothers cannot resist showing off their post-hibernation brood to their human friend. At one point, four generations of matriarchal grizzlies came to visit. All that time, and all those bears, lots and lots of bears, and no single incident threatened or caused human harm. Then, one day in 2008, Alaska Fish and Game officials descended in helicopters, unannounced, geared with fully automatic weapons and flak jackets to confiscate Vandergaw's computer, photos, camera, and other items, and charge him with the illegal feeding of bears. After a prolonged investigation, Vandergaw was charged and received a $20,000 fine and a suspended jail sentence because he gave bears the same food that hunters use to bait bears. His crime was that he had no intent to kill them.[90] Ironically, feeding bears in Alaska is not legal, but baiting them is.[91]

Feeding is defined by law as "placing food materials out that attract wildlife for any reason other than baiting." Baiting also entails putting out food materials to attract wildlife," but with the intent to "lure, or entice them as an aid in hunting."[92] When the bear, deer, or other animal approaches, the hunter who is occupying a tree or manmade "hide" can readily aim and fire without being seen. Hunting aficionados scoff at baiters, maintaining that "real men hunt bears at eye level, not perched in trees swatting mosquitoes and listening to Harlequin romances via audiobook to avoid boredom."[93]

Unperturbed by such judgments, Alaska's Department of Fish and Game enthusiastically endorses the practice and offers free online bear-baiting courses for those sixteen years or older. Its website provides an inventory of bear-baiting tips to aid hunter success, including how and where to build your hide, the best bait, where to shoot on the bear's body, and ways to "reduce the possibility of bears avoiding the site" such as "wearing rubber boots, rubber gloves, and scent shields. Some even go so far as to use scent-free soaps." The techniques listed under "bear baiting tips" constitute "the best form of wildlife management you can practice."[94]

Non-hunting constituencies who may be open to a more welcoming view of bear-human interactions are warned on the Alaska Fish and Game "Living in Harmony with Bears" webpage that "keeping bears away from human food is perhaps the most important thing we can do to prevent conflicts and confrontations between bears and people . . . It is against the law to feed bears. The law states, 'A person may not intentionally feed a moose, deer, elk, bear, wolf, coyote, fox, or wolverine, or negligently leave human food, animal food, or garbage in a manner that attracts these animals.'"[95] In case readers fail to take seriously the admonition that a "fed bear is a dead bear," Fish and Game officials include an inset photo of four bears dead in the back of what we can assume to be an agency truck, a mother bear (referred to as a "sow") and her "three cubs, destroyed as the direct result of improper garbage storage and disposal."[96] Thus, technically speaking, Charlie Vandergaw's crime had nothing to do with supplying food to bears, which is what hunters do routinely with approbation from wildlife agencies. Vandergaw was found guilty with the intent of committing love.

With daily news of mass extinctions and ecological collapse, politicians are beginning to grow weary of Wildlife Services' rogue-style management.[97] After fighting a five-year legal battle to avoid releasing grizzly kill data, it was revealed that British Columbia's provincial government's grizzly numbers had been tampered with, exaggerated to show more bears than actually were counted, and that one of its biologists, Dionys de Leeuw, had been suspended "for suggesting the hunt was excessive and could be pushing the bears to extinction."[98] In 2013, congressmen reprimanded the USDA Wildlife Services unit when it was discovered that the service kills over four million animals annually, give or take a million. Reasons for the disposal of wolves, golden eagles, coyotes, bobcats, and other carnivores were not forthcoming nor available upon request, which prompted Oregon representative Peter DeFazio to refer to Wildlife Services as "one of the most opaque and obstinate departments I've dealt with . . . We're really not sure what they're doing."[99] As an example, in 2012, when Western Regional Director of USDA Wildlife Services Jeffrey Green was asked by the American Society of Mammalogists for basic information regarding its predator control program, he responded, "Just to give tons of raw data to people would not be smart. Torture numbers long enough and they are going to confess to anything."[100] Additional disclosures reveal a pattern. In 2012, "Wildlife Services mistakenly caught and killed more than 520 animals in leghold traps and more than 850 in neck snares, including mountain lions, river otters, pronghorn antelope, deer, badgers, beavers, turtles, turkeys, ravens, ducks, geese, great blue herons, and even a golden eagle."[101]

Eventually data were unearthed and the agency was questioned about its "80 percent" rate of using nonlethal methods. It was revealed that instead, 98 percent of wolves, coyotes, bears, cougars, and other predators that the agency encounters are disposed of by lethal means, that is, killed. The agency posted this disclaimer on their website for years, until 2011:

> NOTICE
>
> Wildlife Services reserves the right to revise the current or any past Program Data Reports to reflect discovery of discrepancies in data presented. Notice of such revisions and their dates will be made on this site when revised versions are posted.

Successive years of data are not posted.[102]

The practice of hiding data and providing misinformation is not limited to the subject of carnivores. APHIS (which is under the umbrella of the U.S. Department of Agriculture) has oversight responsibility for the Fish and Wildlife Service and has been the subject of investigations on the mass deaths of songbirds. News of hundreds upon hundreds of starlings "falling from the sky" remained a mystery until the U.S. Department of Agriculture finally admitted that the cause was poison administered under the auspices of the U.S. government. To this day, the annual deaths of the protected red-winged blackbirds who fall from the sky in Arkansas are dismissed as a natural phenomenon or hominid paranoia.[103] Yet the proffered explanation for blackbird deaths in the millions, New Year's fireworks that spooked the birds, causing some to crash into buildings and others into power lines, has raised eyebrows, especially in light of the fact that Wildlife Services considers that "there is an overpopulation of blackbirds in Beebe and the birds are a potential health hazard" and has undertaken investigations to "reduce blackbird damage" of agricultural crops.[104]

A 2001 USDA report made available on the APHIS website discusses research on how to control blackbirds who were consuming rice crops. (Notably, no mention is made of the appropriation of blackbird habitat and food so critical for sustaining a migratory bird.) Many cited sources in the document are blacked out, even those who are APHIS government employees. The location of two sites where pilot poisoning projects were tested has also been obliterated.[105] These are strange and worrisome omissions in an unclassified government document paid for by public taxpayers.

But, in time, the truth will out. The carnivore myth is growing thin, and the public shows a weariness for killing. Hunting has lost its appeal. A website published by the state of Oregon explains: "The tradition of hunting is on the decline in Oregon and nationwide. Although Oregon's population has grown, real numbers of hunters have declined. In 1980, there were 392,000 resident license holders in Oregon. By 2005, that number had declined to 260,000 licensed hunters."[106] Public opinion increasingly favors carnivore protection, and even some from the opposing camp are changing their minds. The 87 percent of the public in British Columbia who oppose killing grizzlies are "the writing on the wall" that prompted Leonard Ellis to sell his hunting business and reinvest his profits into the Bella Coola Grizzly

Rescued Coyote puppies howling. Photo credit: Jami Hammer.

Tours bear-viewing business. "Change happens whether we like it or not" writes Nancy Macdonald, who documented a controversial grizzly bear hunt in British Columbia.[107] In a level of sensitivity once ridiculed as "bunny-hugging," various news media now display "graphic content" warnings in videos showing animal cruelty such as a fox being torn apart at a British hunt.[108]

When a Canadian conservation officer who refused to kill two orphaned black bear cubs was suspended, over 154,000 people signed a petition in his defense calling for his reinstatement.[109] As Atlantic White Shark Conservancy officials rescued a white shark stranded on a Cape Cod sandbar, beachgoers uncharacteristically cheered and whooped in celebration.[110] Even the forces of law and order are looking less kindly on human violence against wildlife.

In 2014, an Australian man was fined a hefty $18,000 for beating a young white shark to death, and Pope Francis loosed a rosary-rocking encyclical that not so gently chastised science's unrelenting lust for new knowledge.[111] Instead of an endless external search, he urged that attention be directed inward to find solutions that will heal the harm done to nature: "Our goal is not to amass information or to satisfy curiosity, but rather to become painfully aware, to dare to turn what is happening to the world into our own personal suffering and thus to discover what each of us can do about it."[112]

But the old guard is not leaving without a fight, and is unhappy with the change. Decreases in hunting and fishing translate to decreases in revenue. In 2014, the Oregon Department of Fish and Wildlife (ODFW) was faced with a predicted $32 million shortfall because of the drop in hunting and fishing: "sales of state hunting and fishing licenses are at a 30-year low, and operating costs are on the rise."[113] Of the proposed budget of approximately $361 million, "hunters and anglers provide more than forty percent . . . through fees and taxes." Two-thirds of ODFW funding comes from hunting- and fishing-related activities: one-third from state "sale of hunting and fishing licenses," and "another third comes from the federal government—much of it tied to the sale of hunting and fishing equipment." The remaining one-third "comes from a variety of sources," most of which "can be used only for specific purposes spelled out in grants, contracts or statute."[114]

In response, hunting-dominated states are finding new ways to try and counter the drop in interest and concomitant revenues. In Idaho, the Department of Fish and Game offers an efficient online Hunt Planner, "an interactive search and mapping engine" and associated Trip Planner that asks the user: "Planning a hunting trip to Idaho, or just want to see what seasons are open this weekend?" In addition to regulations and hunting tips, the program walks the user through a set of queries on species, weapon type, desired draw odds, and statistics on past kills. Drop-down menus ask users to start by choosing the species they are seeking to kill—wolf, mountain lion, bighorn sheep, bear, deer, and so on. In case you are not too picky or want to increase your odds of bagging some sort of wildlife so you don't go home empty-handed, the program lets you type in more than one species and displays in detail all the various places the animals can best be found. The government agency also offers for eight dollars wolf-trapping classes where participants learn "wolf management; wolf-trapping regulations and ethics; wolf habits and behavior; making, rigging and setting traps and snares; proper care for a wolf; [and] reporting requirements."[115] Listed under the Colorado Parks and Wildlife's website "Things To Do" and wedged between Wildlife Viewing, Family and Kids, and Camping, are three categories of activities that take up the majority of the webpage's space: hunting, fishing, and shooting.[116]

Raising the hunting and fishing fees is not considered a viable option. According to ODFW director Roy Ellicker: "If we were going to raise hunting and fishing license fees to cover our total budget shortfall, that'll drive participation lower, which is not what we want to happen." Instead, the agency has reached out to nontraditional constituencies including women and children.[117] For instance, Oregon has initiated "a new Mentored Youth Hunt program in an effort to get youngsters interested in the sport at an early age. Research demonstrates that early exposure is critical to people taking up hunting or other forms of outdoor recreation. But children and their families are busy today and have many more recreational choices. The current requirement for hunters under the age of 18 to take a hunter education class is a significant time commitment that some may perceive as a barrier to hunting."[118] As an extra incentive, "The program is a 'try before you buy' approach that allows youth ages 9 through 13 to hunt without first passing a hunter education class. It gives unlicensed youngsters the opportunity to receive mentored, one-on-one field training."[119] No hunter education implies no training on safety, which implies kids with lethal weapons on the loose being encouraged to shoot and kill.

Forays into this younger constituency do not sit well with many. Given the relationship between violence against animals and that against humans, wildlife agency efforts are not happy news for law enforcement, teachers, child services, and the public at large. Shaking her head, one New York schoolteacher laments, "Conflating early emotional bonding with killing sets up a time bomb ready to explode. Love and killing—how can anyone with a conscience do that to a child? It is child abuse at its lowest."[120] Statistics concur:

- Acts of animal abuse by children are some of the strongest and earliest diagnostic indicators of conduct disorder, often beginning as young as six-and-a-half years of age
- 70 percent of animal abusers have criminal records, including crimes of violence, property crimes, drug violations, or disorderly behavior
- 50 percent of schoolyard shooters have histories of animal cruelty
- 35 percent of search warrants executed for animal abuse or dog fighting investigations resulted in seizures of narcotics or guns
- 82 percent of offenders arrested for animal-abuse violations had prior arrests for battery, weapons, or drug charges: 23 percent had subsequent arrests for felony offenses
- 70 percent of people charged with cruelty to animals were known by police for other violent behavior—including homicide
- 61.5 percent of animal-abuse offenders had also committed an assault; 17 percent had committed sexual abuse. All sexual homicide offenders reported having been cruel to animals
- 48 percent of rapists and 30 percent of child molesters committed animal abuse in childhood or adolescence.[121]

Meanwhile, the stars of this grisly show, the wildlife, are not faring well. In addition to ever-decreasing habitat and increased human populations, technological advances in weaponry have further stacked the odds against them.[122] Weapon efficacy has gone off the charts with laser scopes, increased firepower, and ATVs that meet and match every hunter's whim. When asked how hunters are able to shoot deer who seem to know with uncanny accuracy the exact days that hunting season starts and ends, a former hunter replied: "Because most hunters these days use scopes that permit sighting in and shooting at a kilometer away."[123] An article in *Outdoor Life* magazine that lists its editors' pick of the ten best deer-hunting rifles sounds more like a jaunty sales advertisement for a car than for a lethal weapon. The Ambush 300 Blackout Hunter is specialized for "small parcels of private ground [where you] need a rifle with the ability to put an animal down quickly. This is a scenario where a large-caliber AR shines. A deer that makes it to land where the hunter doesn't have permission to search can be a nightmare to recover. The Ambush 300 Blackout is well suited for this kind of work. Most hunters will probably opt for the supersonic .30-caliber loads that launch bullets weighing around 115 to 125 grains at 2,200 fps or more. For shots out to 200 yards, these rounds are deadly, and given the moderate recoil and semi-automatic operation of the Ambush, fast follow-up shots are possible if needed."[124]

But what about the fallout from the violence? From the perspective of science, and brain science in particular, is it possible that today's coyote is still a coyote? Although statistics provide numbers, what has the unstoppable carnage over centuries done to the minds of carnivores? Compared to other species, at least in terms of population numbers, the coyote seems able to roll with the genocidal punches with its ability to make do in a variety of habitats, urban or otherwise, and to compensate for untimely deaths in its population by having more pups, more often. It seems that the Trickster has turned the hunters' ways to his own advantage. But, as neuropsychology points out, outward appearances—like stable coyote population numbers—can belie what is happening inside.

Vestigial memories of the carnivore psyche before colonization remain. As Hope Ryden writes, "When what is now Yellowstone National Park was the summer camp of the Crow Indians, the coyote had little to fear from man. His place in the Indian's heart, as well as on his hunting ground, was secure. When the white man began to explore the West, most wildlife, the coyote included, displayed hardly more fear of him than of any other alien species in their midst. Flight distance, or the nearest point to which a man might approach before an animal would flee, was not very great. The Indian's mode of hunting had not instilled in the creatures of the forest and plain the terrible panic reactions they now exhibit at the mere sight of a human silhouette on a far horizon."[125]

The coyote has learned well not to trust the human grim reaper. After a century-and-a-half of indiscriminate killing, coyotes are no longer so numerous, nor do they risk passing "within a few yards of you."[126] The transformation from confident to cowering coyote did not go unnoticed. The Indians "protested to the white man that his method of killing was leaving the surviving animals 'deranged' and impossible to approach . . . many animals were changing not only their habits but their habitats . . . and the trickster coyote had transformed himself into a skulking fugitive. In North America, wildlife had had a taste of civilized man and was seeking hiding places."[127] An Omaha elder's sepia memories testify to the devastation: "When I was a youth, the country was very beautiful. Along the rivers were belts of timberland, where grew cottonwood, maple, elm, ash, hickory, and walnut trees, and many other kinds . . . In both the woodland and the prairie I could see the trails of many kinds of animals and could hear the cheerful songs of many birds. When I walked abroad, I could see many forms of life, beautiful living creatures which Wakanda (the Great Spirit) had placed here; and these were, after their manner, walking, flying, leaping, running, playing all about. But now the face of the land is changed and sad. The living creatures are gone. I see the land desolate and I suffer an unspeakable sadness. Sometimes, I wake in the night, and I feel as though I should suffocate from the pressure of this awful loneliness."[128]

Both carnivores and Indians have changed from the shared experience of genocide. Although the term genocide has been reserved for humans, neuroscience's "species leveling" suggests that we should be seeking a linguistic leveling as well. Genocide, according to the international Convention of the Prevention and Punishment of the Crime of Genocide, is defined as "any of the following acts committed with intent to destroy, in whole or in part, a national, ethnical, racial, or religious group, and includes five types of criminal actions: killing members of the group; causing serious bodily or mental harm to members of the group; deliberately inflicting on the group conditions of life calculated to bring about its physical destruction in whole or in part; imposing measures intended to prevent births within the group; and forcibly transferring children of the group to another group."[129] Coyotes and other carnivores fulfill every criterion. They hold this in common with American Indians. The deadly ripple across time, what Maria Yellow Horse Brave Heart calls "historical trauma," is the legacy of unresolved, intergenerational trauma that has torn the wholeness of pre-colonial prosocial cultures.[130] In place of joy, health, social coherence, and a sense of belonging, the new cultural referents include the inverse: grief, pain, alienation, and dispossession.

Wild carnivores must live like month-to-month renters subject to eviction with no advance notice. They labor under zero tolerance laws that are fueled by the belief that if humans give 'em an indulgent inch, they will take a dangerous mile. When a coyote shows up, a skunk drifts into a backyard, a mother bear and cubs walk down a park trail, or a shark bites a swimmer off the Carolinas, human outrage knows no bounds.[131] The mere hint of their presence brings on an arsenal of traps, poisons, and guns.[132] Any advance into a place of human occupation—even if it is recognized as the rightful address of the animal—is perceived as aggression. Unlucky carnivores who are caught challenging human sovereignty are punished. Nonlethal alternatives may seem to save lives, but they are fraught with difficulties. Translocation mitigation and the use of noisemakers and rubber bullets as "aversive conditioning to teach bears that human spaces are unpleasant" increase stress and mortality.[133] Given the high mortality rates associated with mitigation relocation, many argue that it would be more humane to kill the animals outright.[134]

> **Patient Prognosis** The patient is not likely to fully regain her previous capacity for normal social interactions, nor will she ever be capable of independent living in a safe manner. With continued long-term care (including individual and group psychotherapy), her quality of life can be maintained and enhanced. In the foreseeable future, however, she will continue to present a risk to others and herself due to aggressive outbursts. Although she no longer self-injures, additional trauma or trigger may re-initiate a dysregulated response.

Despite this dark-cloud prognosis for Amber, sanctuary director Jami Hammer sees a brighter lining. "Over the past eleven years since the incident, Amber has improved hugely. She does have bouts of aggression due to the fact that she wants and solicits attention and affection, but just doesn't know how to receive it comfortably without setting her on edge. But, we won't give up. We'll keep on rescuing and helping coyotes as long as they need it. It's been ten years since Amber's shooting. She's fifteen years old now, the oldest coyote at the Indiana rescue center, and we're planning that she stay around for a long time more."[135] Amber died in October 2016 at the age of sixteen. As she was dying, the other sanctuary coyotes began a chorus of howls. They simultaneously fell silent when Amber breathed her last breath.

Epilogue
Pax Carnivora

Dawn opens on silver blue waters. There is not a trace of clouds, and the winds blow a companionable eastward direction, leaving the steady mid-Atlantic groundswell calm and smooth. The diver stands looking out across the blue expanse, studying the slow rise and fall of the seamless ocean waters even as he reaches up then bends back to stretch. It is a really good day, perfect for meeting up with sperm whales: "We suit up on shore because there's no time on board. You have to be ready to jump in immediately when

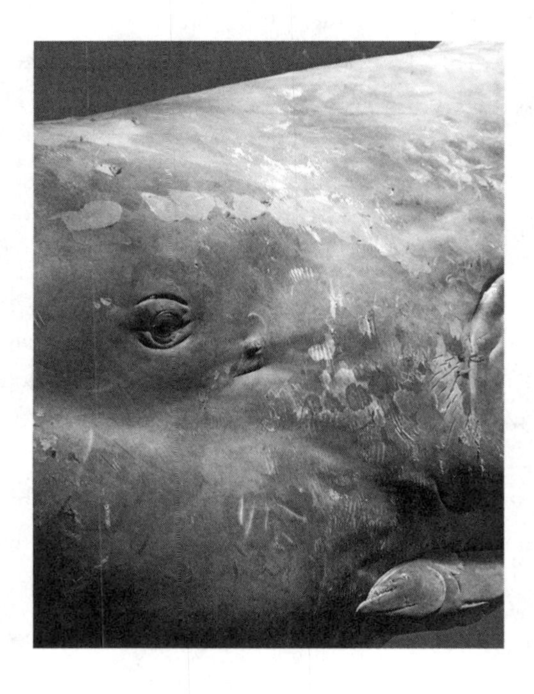

you spot the whales. Sometimes, you end up sitting in your wetsuit for twelve to fourteen hours. That's just what it can take. You have to be patient."[1] Buyle pauses, a smile spreading on his face as recalls the memory.

> That day, we saw a lot of sperm whales, but they were on the move eating, diving down deep to feed then coming up to surface to breathe. They can dive as deep as three thousand or more feet. Amazing. The pod that we were watching wasn't interested in us because they were focused on eating. We stayed in the Zodiac near, not too close so as not to bother them, but close enough to keep track, and watch them disappear then rise to the surface.
>
> After a few hours, we suddenly spotted a group gathered together at the surface. We moved towards them slowly until we were just a hundred meters away. Of course they could hear and see us, but we did not want to invade. We stopped the boat, then carefully slipped into the water and swam toward the group until there was only about thirty meters between us. It was then that I realized something special was going on.
>
> Usually, when sperm whales are gathered to socialize, they are resting and talking to each other, hanging vertically mid-water column. But in this case, they were very active, moving very slowly and deliberately, rolling around each other, and very focused. Then I saw what, *who,* everyone was circling around—it was a tiny baby whale! Obviously, he had been born only minutes before. I could see blood and tissue floating in the water and his tail was still molded and wrapped around him. There was still a piece of the umbilical cord attached. The baby could not even swim on his own. His mother had to push him up to the surface to breathe.
>
> My first reaction was, "Whoa, better be careful!" This is when the group would be feeling the most vulnerable and the mother,

(Overleaf) The benevolent Sperm Whale. Photo credit: Fred Buyle.

> protective. But, there was no sense of tension or hostility at all. Everyone was so soft and gentle. It was incredible . . . You know that all of the adult whales here, outside the Azores, were alive when whaling was still going on? All of them witnessed the slaughter of their families and were hunted down themselves. It must have been terrible—the water full of blood and dead bodies and screaming, dying whales. But the sperm whales don't seem to hold humans any grudge. Here they were, welcoming us to the most sacred of events—the birth of a new baby whale. They welcomed *us,* the same species from the same island of those who had killed them less than thirty years before. That is why I know they are so much more intelligent, so much more sophisticated, so beyond us humans.
>
> Then, I saw that there was a kind of ritual going on. Between breathing times, the mother would push the baby up to meet each member of the family. Up, then over, back to breathe, and on to the next relation—before we knew it, we were surrounded by more than thirty whales! We were so mesmerized that we did not notice that the whales had moved around so that we were positioned right in the middle of them all. Then something extraordinary happened.
>
> The mother started to push the baby in my direction. I just couldn't believe it! All the whales were watching. It seemed like they were smiling at each other and at me. I was so awed. There the mother was accepting us, inviting us humans to join in the family celebration. She knew we were not there to harm, she wanted us included, wanted us to participate in the joy of the newborn. You could feel it. I had worked with them for years, *for years*—but this experience was unbelievable. I had never experienced anything like this before—and never did thereafter. It was mind blowing.[2]

As I came to a close writing this book, I realized that, intended or not, it had to be written. When I completed *Elephants on the Edge* I thought that

Resting Sperm Whale and nearby diver. Photo credit: Fred Buyle.

what was needed to be said, had been said. Once the relationship between cause (human violence) and effect (wildlife psychological breakdown) had been scientifically established, then the path forward to an accepted species-common science of sentience seemed straightforward. The solution also seemed obvious. In words used by mental health professionals—to prevent trauma, you don't cause it—we need simply to stop killing wildlife, stop their capture and captivity, reinstate their right of self-determination, and learn how to live with carnivores the way that Charlie Russell, Fred Buyle, Melissa Amarello, Lawrence Henriques, and the 99 percent of our ancestral small band gathering societies have. If they could do it, then modern humanity with all its wherewithal and can-do attitude certainly is able to accomplish the task. When it comes to human psychological and cultural change, however, the space between theory and practice is of Grand Canyon proportions.

Only a few months after *Elephants on the Edge* was released, South Africa, the very place where elephant PTSD was first diagnosed, reactivated its policy of elephant mass killing. With a stranger-than-fiction nod to the science of elephant trauma, researchers reassuringly stated that new culls would take out entire families—mother, aunties, and babies—to avoid leaving traumatized young elephants in the wake.

The status of the elephants became increasingly dire, as did that of other wildlife species. Scientists began to speak out more openly about impending extinctions and ecosystem collapse. Yet it took several more years before American government and research panels agreed that our closest genetic relatives with whom we share so many acknowledged cultural, social, and psychological qualities, great apes, deserve to be spared further biomedical experimentation—a practice that is prohibited for humans by the Geneva Convention. In this way, the stories of the puma, bear, rattlesnake, orca, crocodile, coyote, and white shark are no different from that of the elephant or those of human tribal peoples who have also suffered genocide and cultural obliteration.

The struggle has just begun. Despite scientists' knowledge of human-animal comparability, the cats whose brains are implanted with electrodes and rats whose bodies are infused with deadly pharmaceuticals remain locked behind sealed doors and walls. Even though science, as modernity's epistemic authority, has proven a thousand times over that nonhuman animals possess the same qualities that grant humans protection, practical implications for this understanding have largely failed to materialize. Lessons from neuropsychology, epigenetics, ethology, and traumatology have been lost.

The main purpose of this book has been to close the chapter on speculation about whether or not animals can feel, think, and suffer, and to move our own species to action—to compel scientists and non-scientists alike to openly accept that we are kin with the finned, feathered, furred, and scaled. Once the distorting lens of modernity is set aside, any outward differences between herbivore and carnivore fade in the glaring light of similarity. Like the elephant, the animal eater can be flooded with emotions. Like the elephant, the puma shares tender moments with friends and family, and like the elephant, the crocodile exercises keen intelligence and restraint, and basks in nature's caressing sunlight. And, like those of the elephants, carnivore minds will, when tormented, contract into the oblivion of inescapable psychic pain for which no amount of learning and evolution can prepare. Their souls are as susceptible to betrayal as any human victim's.

Science has given us a great gift by validating through reason what we know in our hearts. In navigating the madness of wildlife trauma and killing, no theoretical or evolutionary leaps of faith are necessary because neuroscience's predictions match up so well with field observations. Epigenetic change and psychophysiological trauma are the logical effects of radical ecological and social chaos. On this point, evolutionary theory and neuroscience concur: what happens on the outside makes its way to the inside, and vice versa.

Now, at the moment of species re-unification, we have the chance not only to meet the toothed and clawed in the light of reality, but the opportunity to vanquish the self-imposed fear and ignorance that have held humanity hostage. Seeing the world through the eyes of the carnivore is the first step in learning how to relate with these kindred beings, and ourselves, honestly and responsibly. They, and science, teach three key lessons.

The first is that although carnivores wield powerful weapons, they use them with restraint and parsimony. As any veteran warrior attests, hostile engagement is risky and, unlike the padded existence of modernity, life in the wild does not encourage excess of any kind. When the hunter oversteps, something in nature has gone awry. In the case of wildlife, the wrench in nature's otherwise Swiss clockworks is the advent of the Anthropocene. Outbursts of puma, orca, and dolphin internecine killing are not evidence of serial killing instincts, but tragic symptoms of a species desperate to find ground in a world gone mad: as one Vietnam veteran describes it, "PTSD experiences are so traumatic, that we don't ever want to feel again. So one of the things we do to prevent that is to avoid the stressors, which is we withdraw from life—and that's a terrible place to be, really. To have something wonderful to happen and you feel nothing, or something terrible happens—you feel nothing."[3] Within that nothingness, an anythingness descends—P01 murders his mate and kittens; Hugo commits suicide; and Tilikum breaks ancient orca taboos by killing humans.

The second lesson is corollary to the first. Carnivore attacks on humans are the least of our worries. Incidents are so rare as to be statistically insignificant, and nearly all fatal encounters may be avoided. Furthermore, shifting the blame to carnivores for human foibles has left us vulnerable to the real danger: the predator within. The carnivore smear campaign has not deterred the reckoning, only deferred it. On the eve of the Sixth Great Extinction, humanity finds itself in that paradoxical state of getting what it wished for: unchallenged control over the animal kingdom may seem like a glorious achievement to some, but victory is pyrrhic.

The third lesson, germane to Fred Buyle's friend, the sperm whale (*Physeter microcephalus*), is the most important. These mighty whales of the deepest deep are the largest predator on the planet. Great blue whales are larger, but do not possess the sperm whale's ivory teeth. Even if they did not have teeth, the sperm whale's massive body is a mighty weapon. A flick of her or his tail can easily extinguish almost any organism. But the whale does not. The whale refrains. Herein the lesson: power is not synonymous with violence.

Despite what humans have done, sperm whale culture remains inclusive, rooted in the belief of mutual respect. For sperm whales, the counterpoint of victimhood, revenge, is obviously unnecessary, and so it has failed to find a place within their society. Perhaps their seemingly infinite compassion comes from the vast oceans where they live. There space and time, subject and object, fuse in the sperm whale click-and-coda language that envelops the planet like an invisible connective tissue. It is in this gentle world of the whales that we find an ethical exemplar who should inspire our own species' evolution.

To change course now, after such a long history of misunderstanding, will be extraordinarily difficult. Now that the truth has been revealed, we are faced with the enormity of what has been wrought, and left standing at an uncertain crossing. The carnivore crisis forces a very serious choice. Either humans will have to change their attitudes and actions toward carnivores, or we all, like the albatross and dodo, will experience extinction, and soon.

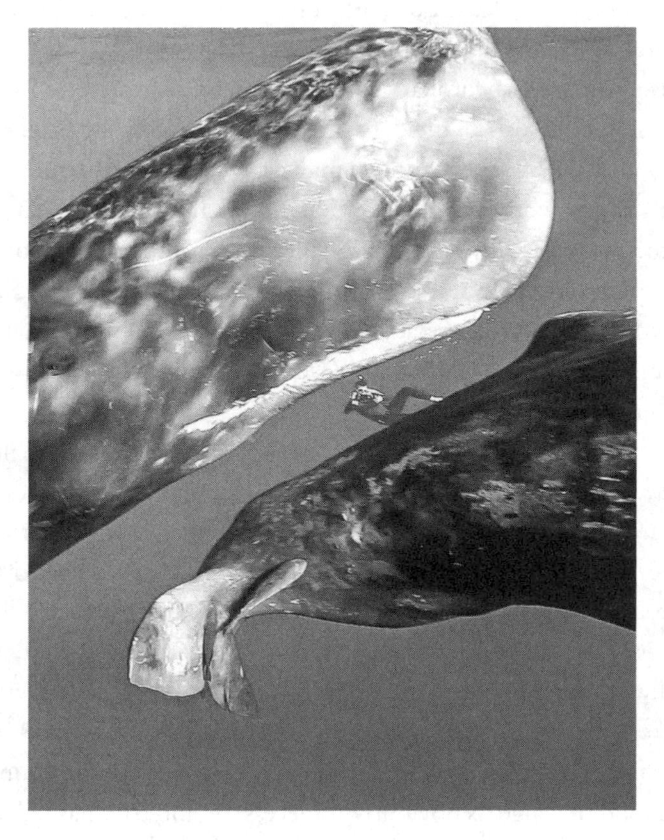

Sperm Whale family off the Azores. Photo credit: Fred Buyle.

Can carnivores be pulled back from the abyss of extinction? Yes. How can their ways of life be restored? The answer is by reversing the values, attitudes, and practices that have driven these species to desperation and annihilation. Simply stop the killing and give back the food and habitat that are the animals' rightful heritage. Tall order? Not when the exigencies of overpopulation, water shortages, pollution, and global warming are considered. Certainly shifting from a millennial-old paradigm to something quite different and of uncertain form will not be easy, nor comfortable. Payback for overdrawing nature's account does not come cheap. But the changes necessary to save carnivores are the same that our species must make to save itself. And so we find that we have no choice but to carry on, but along a different path, one that puts us in harmony with the toothed and clawed—and, as our trans-species ambassadors have shown, what joy this reunion brings!

Meanwhile, in the quiet, where are the whales? In the depths, where they continue to voyage along the Earth's blue skin, singing souls back together in the eternal lyrics of love.

Notes

Foreword

1. P. F. MacNeilage, L. J. Rogers, and G. Vallortigara, "Origins of the Left and Right Brain," *Scientific American* 301 (2009): 60–67; L. J. Rogers, "Relevance of Brain and Behavioural Lateralization to Animal Welfare," *Applied Animal Behavioural Science* 127 (2010): 1–11.

2. See A. N. Schore, *Affect Regulation and the Repair of the Self* (New York: Norton, 2003).

3. See A. N. Schore, *The Science of the Art of Psychotherapy* (New York: Norton, 2012).

Preface

1. Albert Memmi, *The Colonizer and the Colonized* (New York: Routledge, 2013), 129.

Introduction

1. Jesús A. Rivas and Gordon M. Burghardt, "Understanding Sexual Size Dimorphism in Snakes: Wearing the Snake's Shoes," *Animal Behaviour* 62, no. 3 (2001): F1–F6; William Timberlake and Andrew R. Delamater, "Humility, Science, and Ethological Behaviorism," *Behavior Analyst* 14, no. 1 (1991): 37.

2. Harley Shaw, in discussion with the author, June 6, 2015; Harley Shaw, *Soul among Lions: The Cougar as Peaceful Adversary* (Tucson: University of Arizona Press, 2000), 87.

3. Amy Victoria Smith et al., "Functionally Relevant Responses to Human Facial Expressions of Emotion in the Domestic Horse (*Equus caballus*)," *Biology Letters* 12, no. 2 (2016); G. A. Bradshaw, *Elephants on the Edge: What Animals Teach Us about Humanity* (New Haven: Yale University Press, 2009); Knud A. Jønsson, Pierre-Henri Fabre, and Martin Irestedt, "Brains, Tools, Innovation and Biogeography in Crows

and Ravens," *BMC Evolutionary Biology* 12, no. 1 (2012): 1; R. Pronk, D. R. Wilson, and R. Harcourt, "Video Playback Demonstrates Episodic Personality in the Gloomy Octopus," *Journal of Experimental Biology* 213, no. 7 (2010): 1035–1041.

4. Charles Darwin, *The Descent of Man, and Selection in Relation to Sex* (London: J. Murray, 1871), 128.

5. Philip Low et al., "The Cambridge Declaration on Consciousness," *Francis Crick Memorial Conference, 2012: Consciousness in Animals,* http://fcmconference.org/img/CambridgeDeclarationOnConsciousness.pdf, accessed August 20, 2015; Andrew B. Barron and Colin Klein, "What Insects Can Tell Us about the Origins of Consciousness," *Proceedings of the National Academy of Sciences* 113, no. 18 (2016): 4900–4908.

6. Baroness Susan Greenfield CBE, *The Neuroscience of Consciousness,* November 27, 2012, University of Melbourne, https://www.youtube.com/watch?v=k_ZTNmkIiBc, accessed March 12, 2016.

7. *The Standard Edition of the Complete Psychological Works of Sigmund Freud,* vol. 17 (London: Vintage, 2001), 140.

8. Jaak Panksepp and Georg Northoff, "The Trans-Species Core SELF: The Emergence of Active Cultural and Neuro-Ecological Agents through Self-Related Processing within Subcortical-Cortical Midline Networks," *Consciousness and Cognition* 18, no. 1 (2009): 193–215; "Trans-species Psychology," Wikipedia, https://en.wikipedia.org/wiki/Trans-species_psychology, accessed July 17, 2016; G. A. Bradshaw, "Elephant Trauma and Recovery: From Human Violence to Trans-Species Psychology," Ph.D. diss., Pacifica Graduate Institute, Santa Barbara, CA, 2005; Bradshaw, *Elephants on the Edge.*

9. Donald R. Griffin, *Animal Minds: Beyond Cognition to Consciousness* (1992; Chicago: University of Chicago Press, 2013); George Schaller is quoted in "Honouring the Father of Lions: George Adamson," *Zuru Kenya,* 1991, http://zurukenya.com/2013/09/16/honouring-the-father-of-lions-george-adamson, accessed September 16, 2013; Frans De Waal, "Are We in Antropodenial?," *Discover* 18, no. 7 (1997): 50–53; Gordon M. Burghardt, "Cognitive Ethology and Critical Anthropomorphism: A Snake with Two Heads and Hognose Snakes that Play Dead," in C. A. Ristau, ed., *Cognitive Ethology: The Minds of Other Animals* (Hillsdale, NJ: Lawrence Erlbaum, 1991), 53–90.

10. Robert W. Mitchell, Nicholas S. Thompson, and H. Lyn Miles, eds. *Anthropomorphism, Anecdotes, and Animals* (Albany: SUNY Press, 1997); Griffin, *Animal Minds.*

11. Erich D. Jarvis et al., "Opinion: Avian Brains and a New Understanding of Vertebrate Brain Evolution," *Nature Reviews Neuroscience* 6, no. 2 (February 2005): 151–159; G. A. Bradshaw and Barbara L. Finlay, "Natural Symmetry," *Nature* 435, no. 7039 (2005): 149.

12. John Wilson Taylor, "The Athenian Ephebic Oath," *Classical Journal* 13, no. 7 (1918): 495–501.

13. Jessica Bell, "Hierarchy, Intrusion and the Anthropomorphism of Nature: Hunter and Rancher Discourse on North American Wolves," in P. Masius and J. Sprenger, eds., *Historical Interactions between Humans and Wolves* (Isle of Harris, Scotland: White Horse Press, 2015).

14. John Shivik, *The Predator Paradox: Ending the War with Wolves, Bears, Cougars, and Coyotes* (Boston: Beacon Press, 2014).

15. S. G. Platt et al., "Frugivory and Seed Dispersal by Crocodilians: An Overlooked Form of Saurochory?," *Journal of Zoology* 291, no. 2 (2013): 87–99.

16. Michael Mossjan, "U.S. Research Lab Lets Livestock Suffer in Quest for Profit," *New York Times,* January, 19, 2015, http://www.nytimes.com/2015/01/20/dining/animal-welfare-at-risk-in-experiments-for-meat-industry.html, accessed June 27, 2016.

17. John Johnson, "Australia Will Now Kill Sharks on Sight," *Newser,* September 27, 2013, http://www.newser.com/story/154860/australia-will-now-kill-sharks-on-sight.html, accessed June 27, 2016.

18. Victoria Braithwaite, *Do Fish Feel Pain?* (Oxford, Eng.: Oxford University Press, 2010); Forschungsverbund Berlin e.V. (FVB), "Do Fish Feel Pain? Not as Humans Do, Study Suggests," *ScienceDaily,* August 8, 2013, http://www.sciencedaily.com/releases/2013/08/130808123719.htm, accessed June 27, 2016.

19. Ibid.

20. Gay A. Bradshaw and Robert M. Sapolsky, "Macroscope: Mirror, Mirror," *American Scientist* 94, no. 6 (2006): 487–489.

21. J. D. Rose et al., "Can Fish Really Feel Pain?," *Fish and Fisheries* 15, no. 1 (2014): 97–133; Braithwaite, *Do Fish Feel Pain?,* 8; Philip Low et al., "The Cambridge Declaration on Consciousness," *Francis Crick Memorial Conference, 2012: Consciousness in Animals,* http://fcmconference.org/img/CambridgeDeclarationOnConsciousness.pdf, accessed August 20, 2015; Erich D. Jarvis et al., "Opinion: Avian Brains and a New Understanding of Vertebrate Brain Evolution," *Nature Reviews Neuroscience* 6, no. 2 (February 2005): 151–159; Barron and Klein "What Insects Can Tell Us," 4900–4908.

22. Redouan Bshary, Simon Gingins, and Alexander L. Vail, "Social Cognition in Fishes," *Trends in Cognitive Sciences* 18, no. 9 (September 2014): 465–471.

23. Alison Abbott, "Animal Behaviour: Inside the Cunning, Caring, and Greedy Minds of Fish," *Nature* 521, no. 7553 (May 26, 2015): 412–414, http://www.nature.com/news/animal-behaviour-inside-the-cunning-caring-and-greedy-minds-of-fish-1.17614, accessed June 27, 2016.

24. Rusty Barnes, Charlie Tuna Starkist Commercial, 2014, https://www.youtube.com/watch?v=LktyRYmDyPc, accessed June 27, 2016.

25. Maurice G. Hornocker, *Cougar Ecology and Conservation* (Chicago: University of Chicago Press, 2010), 85.

26. Shivik, *Predator Paradox,* 61.

27. Lisa Naughton-Treves, Rebecca Grossberg, and Adrian Treves, "Paying for Tolerance: Rural Citizens' Attitudes toward Wolf Depredation and Compensation," in *Conservation Biology* 17, no. 6: 1500–1511, 1508; Bell, "Hierarchy, Intrusion and the Anthropomorphism of Nature."

28. Bell, "Hierarchy, Intrusion and the Anthropomorphism of Nature."

29. Daphne Sheldrick, *The Tsavo Story* (London: Harvill Press, 1973).

30. Shivik, *Predator Paradox,* 9; Kevin Hansen and Robert Redford, *Cougar: The American Lion* (Flagstaff, AZ: Northland Publications, 1992), 59.

31. "Tribes Could Kill Columbia River Sea Lions under New Bill: Oregon Environment Roundup," *OregonLive.com,* http://www.oregonlive.com/environment/index.ssf/2015/01/tribal_members_could_kill_colu.html, accessed August 19, 2015; "Govt to Cull 10,000 Cormorants to Protect Columbia River Salmon," *Al Jazeera America,* http://america.aljazeera.com/articles/2015/5/28/cormorants-culled-to-protect-columbia-river-salmon.html, accessed June 19, 2015.

32. U.S. Fish and Wildlife Service, "2011 National Survey of Fishing, Hunting, and Wildlife-Associated Recreation: National Overview," accessed August 19, 2015, http://digitalmedia.fws.gov/cdm/ref/collection/document/id/859; National Shooting Sports Foundation, "Hunting in America: An Economic Force for Conservation," http://www.nssf.org/hunting, accessed July 1, 2015.

33. Economists at Large, *The $200 Million Question: How Much Does Trophy Hunting Really Contribute to African Communities?,* a report for the African Lion Coalition by Economists at Large, Australia, http://www.ifaw.org/sites/default/files/Ecolarge-2013-200m-question.pdf, accessed August 3, 2015; Jane Mayer, "Does Zimbabwe Really Need Trophy Hunting?," *New Yorker,* http://www.newyorker.com/news/news-desk/does-zimbabwe-really-need-trophy-hunting, accessed August 2, 2015; M. E. Smith and D. Molde, Nevadans for Responsible Wildlife Management, *Wildlife Conservation and Management Funding in the U.S.,* 2014, http://www.mountainlion.org/featureimages/whopaysforwildlife/USA-O-NRWM-Smith-Molde-2014-Wildlife-Conservation-Management-Funding-in-the-US.pdf, accessed July 18, 2016.

34. "Home Injuries Rising, Often Deadly," WebMD, http://www.webmd.com/healthy-aging/news/20021002/home-injuries-rising-often-deadly, accessed August 19, 2015.

35. Federal Bureau of Investigation, *2014 Crime in the United States,* https://www.fbi.gov/news/stories/2015/september/latest-crime-stats-released/latest-crime-stats-released, accessed March 2016.

36. John Muir, *Our National Parks* (Boston: Houghton Mifflin, 1901).

37. Julie Gilchrist, J. J. Sacks, D. White, and M. J. Kresnow, "Dog Bites: Still a Problem? "*Injury Prevention* 14, no. 5 (2008): 296–301; "Dog Attack Deaths and Maimings, U.S. and Canada September 1982 to December 31, 2014, Reports," http://www.dogsbite.org/dog-bite-statistics-study-dog-attacks-and-maimings-merritt-clifton.php, accessed June 27, 2016; Jeffrey J. Sacks, Leslie Sinclair, and Julie Gilchrist, "Breeds of Dogs Involved in Fatal Human Attacks in the United States between 1979 and 1998," *Journal of the American Veterinary Medical Association* 217, no. 6 (2000): 836–840; Shivik, *Predator Paradox,* 62; "Lion Population Number Declines—Problem Animal Control or Trophy Hunting?," *LionAid,* last updated December 28, 2011, http://www.lionaid.org/news/2011/12/lion-population-number-declines-problem-animal-control-or-trophy-hunting.htm, accessed July 6, 2015.

38. Val Plumwood, *The Eye of the Crocodile,* ed. Lorraine Shannon (Canberra: ANU EPress, 2013).

39. James Hatley, "The Uncanny Goodness of Being Edible to Bears," in *Rethinking Nature: Essays in Environmental Philosophy,* eds. Bruce V. Foltz and Robert Frodeman (Bloomington: Indiana University Press, 2004), 21.

40. Bernd Brunner, *Bears: A Brief History,* trans. Lori Lantz (New Haven: Yale University Press, 2008).

41. Shivik, *Predator Paradox,* 4.

42. Hansen and Redford, *Cougar,* 56.

43. Barry Lopez, *Of Wolves and Men* (New York: Charles Scribner's Sons, 1978); Hansen and Redford, *Cougar* 57.

44. U.S. Fish and Wildlife Service, "National Conservation Center," http://training.fws.gov/history/TimelinesOrigins.html, accessed June 27, 2016; U.S. Fish and Wildlife Service, "Who We Are," https://www.fws.gov/who/; accessed July 10, 2016.

45. Kim Murray Berger, "Carnivore-Livestock Conflicts: Effects of Subsidized Predator Control and Economic Correlates on the Sheep Industry," *Conservation Biology* 20, no. 3 (2006): 751–761; William J. Ripple et al., "Widespread Mesopredator Effects after Wolf Extirpation," *Biological Conservation* 160 (2013): 70–79; Hansen and Redford, *Cougar.*

46. Ibid.

47. William J. Ripple et al., "Widespread Mesopredator Effects after Wolf Extirpation," *Biological Conservation* 160 (2013): 70–79.

48. Ronald Tilson and Philip J. Nyhus, eds., *Tigers of the World: The Science, Politics, and Conservation of Panthera tigris* (San Diego: Academic Press, 2009); Panthera, "Panthera," https://www.panthera.org/cat/tiger; accessed July 17, 2016; J. K. Baum et al., "Collapse and Conservation of Shark Populations in the Northwest Atlantic," *Science* 299 (2003): 389–392.

49. Zheng Huan Wang et al., "Testing Reintroduction as a Conservation Strategy

for the Critically Endangered Chinese Alligator: Movements and Home Range of Released Captive Individuals," *Chinese Science Bulletin* 56, no. 24 (2011): 2586–2593.

50. Tom Knudson, "The Killing Agency: Wildlife Services' Brutal Methods Leave a Trail of Animal Death," *Sacramento Bee,* April 28, 2012, http://www.sacbee.com/news/investigations/wildlife-investigation/article2574599.html, accessed February 2016.

51. Ibid.

52. "'Carnivore Cleansing' Is Damaging Ecosystems, Scientists Warn," *Guardian,* January 9, 2014, http://www.theguardian.com/environment/2014/jan/09/carnivore-cleansing-damaging-ecosystems, accessed June 27, 2016.

53. W. J. Ripple et al., "Status and Ecological Effects of the World's Largest Carnivores," *Science* 343, no. 6167 (January 10, 2014): 6164, http://www.sciencemag.org/content/343/6167/1241484, accessed June 27, 2016.

54. "'Carnivore Cleansing.'"

55. Dr. Zohara M. Hieronimus, personal communication with the author, August 8, 2015.

56. Kevin J. Van Tighem, *Bears: Without Fear* (Victoria, B.C.: Rocky Mountain Books, 2013).

57. Gordon M. Burghardt, "Human-Bear Bonding in Research on Black Bear Behavior," in Hank Davis and Dianne A. Balfour, eds., *The Inevitable Bond: Examining Scientist-Animal Interactions* (Cambridge, Eng.: Cambridge University Press, 1992), 365–382.

58. Mark A. Ditmer et al., "Bears Show a Physiological But Limited Behavioral Response to Unmanned Aerial Vehicles," *Current Biology,* http://linkinghub.elsevier.com/retrieve/pii/S0960982215008271, accessed August 21, 2015, doi:10.1016/j.cub.2015.07.024; P. C. Paquet and C. T. Darimont, "Wildlife Conservation and Animal Welfare: Two Sides of the Same Coin," *Animal Welfare* 19, no. 2 (2010): 177–190.

59. Ibid.

60. Bruce D. Perry, "Neurobiological Sequelae of Childhood Trauma: PTSD in Children," in M. Michele Murburg, ed., *Catecholamine Function in Posttraumatic Stress Disorder: Emerging Concepts* (Washington, DC: American Psychiatric Press, 1994), 253–276.

61. J. Moussaieff Masson, *Beasts: What Animals Can Teach Us about the Origins of Good and Evil* (New York: Bloomsbury, 2014).

62. Chris T. Darimont, Caroline H. Fox, Heather M. Bryan, and Thomas E. Reimchen, "The Unique Ecology of Human Predators," *Science* 349, no. 6250 (2015): 858–860.

Chapter 1. White Sharks

1. Frederic Buyle, in discussion with the author, April 17, 2015.

2. Leonard J. V. Compagno, *Sharks of the World: An Annotated and Illustrated Catalogue of Shark Species Known to Date,* vol. 2, no. 1 (New York: Food & Agriculture Org., 2001); World Wildlife Fund, "Sharks," http://www.worldwildlife.org/species/shark, accessed March 16, 2016.

3. Darewin, http://www.darewin.org, accessed May 23, 2015.

4. Ralph Collier, in discussion with the author, March 12, 2016.

5. Compagno, *Sharks of the World.*

6. Tanya Lewis, "Amazing Footage of Sharks Swimming in Scalding Waters around a Volcano Is Completely Baffling Scientists," *Yahoo Finance,* July 10, 2015, http://finance.yahoo.com/news/amazing-footage-sharks-swimming-boiling-155007793.html, accessed August 13, 2015; Carolyn Barnwell, "Deep-Sea Cameras Reveal 'Sharkcano,'" July 9, 2015, http://voices.nationalgeographic.com/2015/07/09/cameras-reveal-sharkcano-sharks-in-volcano, accessed August 22, 2015.

7. Natalie O'Neill, "Huge Great White Shark Swimming toward New York," *New York Post,* May 10, 2015, http://nypost.com/2015/05/10/huge-great-white-shark-swimming-toward-new-york, accessed August 22, 2015.

8. Mary Lee's Facebook page, https://www.facebook.com/Maryleeshark, accessed August 22, 2015.

9. N. C Wegner et al., "Whole-Body Endothermy in a Mesopelagic Fish, the Opah, *Lampris guttatus,*" *Science* 348, no. 6236 (2015): 786–789.

10. Stephen L. Katz, "Design of Heterothermic Muscle in Fish," *Journal of Experimental Biology* 205 (2002): 2251–2266.

11. Douglas J. Long, "Records of White Shark-Bitten Leatherback Sea Turtles along the Central California Coast," in A. Peter Klimley and David G. Ainley, eds., *Great White Sharks: The Biology of Carcharodon carcharias* (San Diego: Academic Press, 1996), 317–319; Alessandro De Maddalena and Walter Heim, *Mediterranean Great White Sharks: A Comprehensive Study Including All Recorded Sightings* (Jefferson, N.C.: McFarland & Co., 2012), 189.

12. Fred Buyle, in discussion with the author, April 17, 2015.

13. Ralph S. Collier, in discussion with the author, March 11, 2016.

14. Jayne M. Gardiner et al., "Multisensory Integration and Behavioral Plasticity in Sharks from Different Ecological Niches," *PLoS ONE* 9, no. 4 (April 2, 2014): e93036, doi:10.1371/journal.pone.0093036.

15. A. Peter Klimley and David G Ainley, eds., *Great White Sharks: The Biology of Carcharodon carcharias* (San Diego: Academic Press, 1996).

16. Conrad W. Speed et al., "Complexities of Coastal Shark Movements and Their Implications for Management," *Marine Ecology Progress Series* 408 (2010): 275–293.

17. Neil Hammerschlag, R. Aidan Martin, and Chris Fallows, "Effects of Environmental Conditions on Predator-Prey Interactions between White Sharks (*Carcharodon carcharias*) and Cape Fur Seals (*Arctocephalus pusillus pusillus*) at Seal Island, South Africa," *Environmental Biology of Fishes* 76, nos. 2–4 (2006): 341–350; Charlie Huveneers et al., "White Sharks Exploit the Sun during Predatory Approaches," *American Naturalist* 185, no. 4 (2015): 562–570.

18. ReefQuest Center for Shark Research, http://www.elasmo-research.org, accessed March 1, 2016.

19. Itsumi Nakamura, Carl G. Meyer, and Katsufumi Sato, "Unexpected Positive Buoyancy in Deep Sea Sharks, *Hexanchus griseus,* and a *Echinorhinus cookei,*" *PLoS ONE* 10, no. 6 (June 10, 2015): e0127667, http://journals.plos.org/plosone/article?id=10.1371/journal.pone.0127667, accessed August 22, 2015.

20. Marie Levine, in discussion with the author, March 12, 2016.

21. Global Shark Attack File, http://www.sharkattackfile.net, accessed March 2016.

22. Marie Levine, in discussion with the author, July 2016.

23. Ralph S. Collier, *Shark Attacks of the Twentieth Century: From the Pacific Coast of North America* (Transylvania: Scientia Publishing, 2003).

24. Collier quoted in Jennifer Hile, "Great White Shark Attacks: Defanging the Myths," *National Geographic,* January 23, 2004; http://news.nationalgeographic.com/news/2004/01/0123_040123_tvgreatwhiteshark.html, accessed November 3, 2015.

25. R. Aiden Martin quoted in Hile, "Great White Shark Attacks."

26. Shark Research Committee, "Unprovoked White Shark Attacks on Surfers," http://www.sharkresearchcommittee.com/unprovoked_surfer.htm, accessed August 11, 2015; Collier, *Shark Attacks of the Twentieth Century.*

27. Ralph S. Collier, in discussion with the author, March 11, 2016.

28. ReefQuest Centre for Shark Research, "Catch as Catch Can," http://www.elasmo-research.org/education/topics/b_catch.htm, accessed August 9, 2015.

29. Sanford A. Moss, *Sharks: An Introduction for the Amateur Naturalist* (New York: Prentice Hall, 1984); Ralph S. Collier, in discussion with the author, March 11, 2016.

30. Ralph S. Collier, in discussion with the author, March 14, 2016.

31. Sora L. Kim et al., "Ontogenetic and Among-Individual Variation in Foraging Strategies of Northeast Pacific White Sharks Based on Stable Isotope Analysis," *PLoS ONE* 7, no. 9 (September 28, 2012): e45068, doi:10.1371/journal.pone.0045068.

32. Ralph S. Collier, in discussion with the author, March 14, 2016.

33. Ibid.

34. ReefQuest Centre for Shark Research, "Catch as Catch Can."

35. G. B. Skomal, E. M. Hoyos-Padilla, A. Kukulya, and R. Stokey, "Subsurface

Observations of White Shark *Carcharodon carcharias* Predatory Behaviour Using an Autonomous Underwater Vehicle," *Journal of Fish Biology* 87, no. 6 (2015): 1293–1312, 1306; Marc Aquino Baleytó, in discussion with the author, March 10, 2016.

36. Stephen Wroe et al., "Three-Dimensional Computer Analysis of White Shark Jaw Mechanics: How Hard Can a Great White Bite?," *Journal of Zoology* 276 (2008): 336–342, http://www.academia.edu/239896/Three-dimensional_computer_analysis_of_white_shark_jaw_mechanics_how_hard_can_a_great_white_bite, accessed July 23, 2015; Gregory M. Erickson et al., "Insights into the Ecology and Evolutionary Success of Crocodilians Revealed through Bite-Force and Tooth-Pressure Experimentation," *PLoS ONE* 7, no. 3 (March 14, 2012): e31781, http://journals.plos.org/plosone/article?id=10.1371/journal.pone.0031781#pone-0031781-t001, accessed July 30, 2015.

37. E. J. Brooks et al., "The Stress Physiology of Extended Duration Tonic Immobility in the Juvenile Lemon Shark, Negaprion Brevirostris (Poey 1868)," *Journal of Experimental Marine Biology and Ecology* 409, nos. 1–2 (2011): 351–360.

38. Bill Blakemore, "Whale Kills Shark, Setting Biology on Its Ear," *ABC News,* February 24, 2010, http://abcnews.go.com/GMA/Weekend/killer-whale-caught-tape-killing-great-white-shark/story?id=9191986, accessed August 4, 2015.

39. Peter Pyle et al., "Predation on a White Shark (*Carcharodon carcharias*) by a Killer Whale (*Orcinus orca*) and a Possible Case of Competitive Displacement," *Marine Mammal Science* 15, no. 2 (1999): 563–568.

40. Mitchell A. Watsky and Samuel H. Gruber, "Induction and Duration of Tonic Immobility in the Lemon Shark, *Negaprion brevirostris,*" *Fish Physiology and Biochemistry* 8, no. 3 (1990): 207–210.

41. Brooks, "Stress Physiology," 357.

42. Ibid., 351–360; Takayoshi Nishida, "Adaptive Significance of Death Feigning Posture as a Specialized Inducible Defense against Gape-Limited Predators," *Proceedings of the Royal Society B: Biological Sciences* 273, no. 1594 (2006): 1631–1636.

43. B. A. Block et al., "Tracking Apex Marine Predator Movements in a Dynamic Ocean," *Nature* 475, no. 7354 (2011): 86–90.

44. Mindy Weisberger, "Surprise! Sharks Have Social Lives," *Science Live,* February 23, 2016.

45. Michael L. Domeier, ed., *Global Perspectives on the Biology and Life History of the White Shark* (Boca Raton, FL: CRC Press, 2012).

46. Emilio Sperone et al., "Social Interactions among Bait-Attracted White Sharks at Dyer Island (South Africa)," *Marine Biology Research* 6, no. 4 (2010): 408–414.

47. David M. P. Jacoby, Darren P. Croft, and David W. Sims, "Social Behaviour in Sharks and Rays: Analysis, Patterns, and Implications for Conservation," *FAF Fish and Fisheries* 13, no. 4 (2012): 399–417.

48. Charles Darwin, Paul Ekman, and Phillip Prodger, *The Expression of the Emotions in Man and Animals* (New York: Oxford University Press, 1998).

49. Danny Clemens, "Learn More about Deep Blue, One of the Biggest Great White Sharks Ever Filmed," *Discovery*, last modified June 10, 2015, http://www.discovery.com/tv-shows/shark-week/shark-feed/learn-more-about-deep-blue-one-of-the-biggest-great-white-sharks-ever-filmed, accessed August 23, 2015.

50. Salvador J. Jorgensen et al., "Connectivity among White Shark Coastal Aggregation Areas in the Northeastern Pacific," in Michael L. Domeier and the International White Shark Symposium, eds., *Global Perspectives on the Biology and Life History of the White Shark* (Boca Raton, FL: CRC Press, 2012): 159–168; Salvador J. Jorgensen et al., "Philopatry and Migration of Pacific White Sharks," *Proceedings of the Royal Society B* 277, no. 1682 (2010): 679–688.

51. R. L. Robbins et al., "Residency and Local Connectivity of White Sharks at Liguanea Island: A Second Aggregation Site in South Australia?" *Open Fish Science Journal* 8, no. 1 (2015); Ralph S. Collier, in discussion with the author, March 11, 2016.

52. Ralph Collier, in discussion with the author, March 11, 2016.

53. Marc Aquino Baleytó, in discussion with the author, March 2016; http://abcnews.go.com/US/spearfisher-captures-gnarly-great-white-shark-attack-camera/story?id=42048744

54. Fred Buyle, in discussion with the author, May 30, 2015.

55. Mauricio Hoyo de Padilla, in discussion with the author, June 10, 2015; R. Aidan Martin, Alessandro De Maddalena, and ReefQuest Centre for Shark Research, *Field Guide to the Great White Shark* (Vancouver: ReefQuest Centre for Shark Research, 2003).

56. Ralph S. Collier, in discussion with the author, March 11, 2016.

57. J. Matthias Starck and Robert E. Ricklefs, *Avian Growth and Development: Evolution within the Altricial-Precocial Spectrum* (New York: Oxford University Press, 1998), 241.

58. Laure Frésard et al., "Epigenetics and Phenotypic Variability: Some Interesting Insights from Birds," *Genetics Selection Evolution* 45, no. 1 (June 11, 2013): 16, http://www.gsejournal.org/content/45/1/16, 7, accessed August 15, 2015.

59. M. J. Meaney, "Maternal Care, Gene Expression, and the Transmission of Individual Differences in Stress Reactivity across Generations," *Annual Review of Neuroscience* 24 (2001): 1161–1192; R. C. Bagot and M. J. Meaney, "Epigenetics and the Biological Basis of Gene X Environment Interactions," *Journal of the American Academy of Child and Adolescent Psychiatry* 49, no. 8 (2010), 767.

60. Michael J. Meaney, "Epigenetics and the Environmental Regulation of the Genome and Its Function," in Darcia Narvaez, ed., *Evolution, Early Experience and Human Development: From Research to Practice and Policy* (Oxford, Eng.: Oxford University Press, 2012), 117.

61. Ibid., 752–771.

62. Robert R. McCrae and Oliver P. John, "An Introduction to the Five-Factor Model and Its Applications." *Journal of Personality* 60, no. 2 (1992): 175–215.

63. Julien F. Ayroles et al., "Behavioral Idiosyncrasy Reveals Genetic Control of Phenotypic Variability," *Proceedings of the National Academy of Sciences* 112, no. 21 (2015): 6706–6711.

64. Samuel D. Gosling, "From Mice to Men: What Can We Learn about Personality from Animal Research?," *Psychological Bulletin* 127, no. 1 (2001): 45.

65. David M. P. Jacoby et al., "Shark Personalities? Repeatability of Social Network Traits in a Widely Distributed Predatory Fish," *Behavioral Ecology and Sociobiology* 68, no. 12 (2014): 1995–2003.

66. Lauren J. N. Brent et al., "Personality Traits in Rhesus Macaques (*Macaca mulatta*) Are Heritable but Do Not Predict Reproductive Output," *International Journal of Primatology* 35, no. 1 (2014): 188–209.

67. Samuel D. Gosling and Simine Vazire, "Are We Barking up the Right Tree? Evaluating a Comparative Approach to Personality," *Journal of Research in Personality* 36, no. 6 (2002): 607–614.

68. Kenneth L. Davis and Jaak Panksepp, "The Brain's Emotional Foundations of Human Personality and the Affective Neuroscience Personality Scales," *Neuroscience and Biobehavioral Reviews* 35, no. 9 (2011): 1946–1958.

69. Fred Buyle, in discussion with the author, May 30, 2015.

70. Ibid.

71. Sophie Jane Evans, "Heavily Pregnant Bethany Hamilton Seen Walking Dogs and Going for Swim," *Mail Online,* May 8, 2015, http://www.dailymail.co.uk/news/article-3074174/Shark-attack-survivor-Bethany-Hamilton-25-shows-huge-baby-bump-walks-beloved-pet-dogs-husband-Adam-going-swim.html, accessed August 23, 2015.

72. Ralph S. Collier, in discussion with the author, March 11, 2016.

73. Fred Buyle, in discussion with the author, May 30, 2015.

74. Ibid.

75. Charlie Russell, in discussion with the author, March 2, 2013.

76. Marie Levine, in discussion with author, May 6, 2016.

77. Fred Buyle, in discussion with the author, May 30, 2015.

78. Fischer quoted in Alec Wilkerson, "Cape Fear," *New Yorker,* September 9, 2013, http://www.newyorker.com/magazine/2013/09/09/cape-fear, accessed August 23.

79. Ibid.

80. Laurence J. Kirmayer, Christopher Fletcher, and Robert Watt, "Locating the Ecocentric Self: Inuit Concepts of Mental Health and Illness," in Laurence J. Kirmayer and Gail Guthrie Valaskakis, eds., *Healing Traditions: The Mental Health of Aboriginal Peoples in Canada* (Vancouver: University of British Columbia Press, 2009), 289–314.

81. Peter Fimrite, "'Porpicide': Bottlenose Dolphins Killing Porpoises," *SFGate,* last updated September 17, 2011, http://www.sfgate.com/news/article/Porpicide-Bottle nose-dolphins-killing-porpoises-2309298.php, accessed August 23, 2015; Nigel Blundell, "Killer Dolphins Baffle Marine Experts," *The Telegraph,* January 25, 2008, http:// www.telegraph.co.uk/news/earth/earthnews/3323070/Killer-dolphins-baffle-marine -experts.html, accessed August 23, 2015.

82. "Californian Dolphin Gang Caught Killing Porpoises," *New Scientist,* last modified June 1, 2011, https://www.newscientist.com/article/mg21028154.700-californian -dolphin-gang-caught-killing-porpoises/#.VWOCRoYkqAU, accessed August 23, 2015.

83. Mark P. Cotter, Daniela Maldini, and Thomas A. Jefferson, "'Porpicide' in California: Killing of Harbor Porpoises (*Phocoena phocoena*) by Coastal Bottlenose Dolphins (*Tursiops truncatus*)," *Marine Mammal Science* 28, no. 1 (2012): 1.

84. "*Tursiops truncatus* (Bottle-Nosed Dolphin, Bottlenosed Dolphin, Bottlenose Dolphin, Common Bottlenose Dolphin)," http://www.iucnredlist.org/details/22563/0, accessed August 3, 2015.

85. IUCN News Report, "A Quarter of Sharks and Rays Threatened with Extinction," January 21, 2014, http://www.iucn.org/?14311/A-quarter-sharks-and-rays -threatened-with-extinction, accessed July 19, 2016; Taylor K. Chapple et al., "A First Estimate of White Shark, Carcharodon *carcharias,* Abundance off Central California," *Biology Letters* (2011): rsbl20110124; George H. Burgess et al., "A Re-Evaluation of the Size of the White Shark (*Carcharodon carcharias*) Population Off California, USA," *PloS One* 9, no. 6 (2014): e98078.

86. Shelley C. Clarke, Shelton J. Harley, Simon D. Hoyle, and Joel S. Rice, "Population Trends in Pacific Oceanic Sharks and the Utility of Regulations on Shark Finning," *Conservation Biology* 27, no. 1 (2013): 197–209.

87. Boris Worm et al., "Global Catches, Exploitation Rates, and Rebuilding Options for Sharks," *Marine Policy* 40 (2013): 194–204; "Shark Conservation in China," Pew Charitable Trust, http://bit.ly/1J6wwIQ, accessed August 2, 2015.

88. NOAA Fisheries, "Marine and Anadromous Fish," http://www.nmfs.noaa.gov/ pr/species/fish, accessed June 27, 2016; NOAA Fisheries, "International Shark Conservation," http://www.nmfs.noaa.gov/ia/species/sharks/shark.html; accessed March 16, 2016.

89. Fred Buyle, in discussion with the author, June 9, 2015.

Chapter 2. Grizzly Bears

1. Deke Weaver, *Polar Bear God,* https://vimeo.com/54963015, accessed November 2015.

2. C. C. Schwartz, S. D. Miller, and M. Haroldson, "Grizzly Bear," in George A.

Feldhamer, Bruce Carlyle Thompson, and Joseph A. Chapman, eds., *Wild Mammals of North America: Biology, Management, and Conservation,* (Baltimore: Johns Hopkins University Press, 2003), 556.

3. Ibid.

4. Ibid.; David J. Mattson, Stephen Herrero, R. Gerald Wright, and Craig M. Pease, "Science and Management of Rocky Mountain Grizzly Bears," *Conservation Biology* 10, no. 4 (1996): 1013–1025.

5. Enos A. Mills, *The Grizzly, Our Greatest Wild Animal* (Boston: Houghton Mifflin Company, 1919), 102.

6. John Muir, *Steep Trails* (Champaign, IL: Project Gutenberg), http://www.gutenberg.org/ebooks/326, accessed June 27, 2016.

7. Lee Stetson, *The Wild Muir: Twenty-Two of John Muir's Greatest Adventures* (San Francisco: Yosemite Conservancy, 2013), 41.

8. Joaquin Miller, *True Bear Stories* (Chicago: Rand, McNally, 1900), http://www.gutenberg.org/ebooks/40869, accessed June 27, 2016.

9. S. Simpson, *Dominion of Bears: Living with Wildlife in Alaska* (Lawrence: University of Kansas Press, 2013).

10. Aldo Leopold, *A Sand County Almanac, and Sketches Here and There* (New York: Oxford University Press, 1987), 136–137.

11. Ibid., 137.

12. Charlie Russell, in discussion with the author, April 6, 2013.

13. Mark J. Palmer, "Bring Back the California Grizzly," *Counterpunch,* last modified November 8, 2002, http://www.counterpunch.org/2002/11/08/bring-back-the-california-grizzly, accessed June 27, 2016.

14. Charlie Vandergaw, in discussion with the author, June 8, 2014.

15. Charlie Russell, *Grizzly Heart* (Toronto: Random House of Canada, 2011).

16. Matt W. Hayward and Michael Somers, eds., *Reintroduction of Top-Order Predators* (New York: John Wiley & Sons, 2009); Scott R. Derrickson et al., *Status of the California Condor and Efforts to Achieve Its Recovery* (New York: American Ornithologists' Union, 2008).

17. G. A. Bradshaw and A. N. Schore, "How Elephants Are Opening Doors: Developmental Neuroethology, Attachment, and Social Context," *Ethology* 113 (2007): 426–436.

18. A. S. Fleming, D. H. O'Day, and G. W. Kraemer, "Neurobiology of Mother-Infant Interactions: Experience and Central Nervous System Plasticity across Development and Generations," *Neuroscience Biobehavioral Reviews* 23 (1999): 673–685; G. A. Bradshaw and A. N. Schore, "How Elephants Are Opening Doors: Developmental Neuroethology, Attachment, and Social Context," *Ethology* 113 (2007): 426–436; Georg Northoff and Jaak Panksepp, "The Trans-Species Concept of Self and the

Subcortical-Cortical Midline System," *Trends in Cognitive Sciences* 12, no. 7 (2008): 259–264; Jaak Panksepp and Georg Northoff, "The Trans-Species Core SELF: The Emergence of Active Cultural and Neuro-Ecological Agents through Self-Related Processing within Subcortical-Cortical Midline Networks," *Consciousness and Cognition* 18, no. 1 (2009): 193–215.

19. Darcia Narvaez, *Embodied Morality: Protectionism, Engagement and Imagination* (New York: Palgrave MacMillan), in press; Allan N. Schore, *Affect Regulation and the Origin of the Self: The Neurobiology of Emotional Development* (1994; New York: Routledge, 2015).

20. Darcia Narvaez, *Neurobiology and the Development of Human Morality: Evolution, Culture, and Wisdom* (New York: W. W. Norton, 2014).

21. John Bowlby, *Attachment,* vols. 1–3 of Bowlby, *Attachment and Loss* (New York: Basic Books, 1969).

22. Schore, *Affect Regulation and the Origin of the Self;* Allan N. Schore, *Affect Dysregulation and Disorders of the Self (Norton Series on Interpersonal Neurobiology)* (New York: W.W. Norton, 2003); Allan N. Schore, *Affect Regulation and the Repair of the Self (Norton Series on Interpersonal Neurobiology),* vol. 2 (New York: W.W. Norton, 2003).

23. Allan N. Schore, "Effects of a Secure Attachment Relationship on Right Brain Development, Affect Regulation, and Infant Mental Health," *Infant Mental Health Journal* 22 (2001): 7–66.

24. Ibid., 23.

25. Allan N. Schore, "Attachment and the Regulation of the Right Brain," *Attachment & Human Development* 2, no. 1 (2000): 23–47, 23.

26. Edward Z. Tronick, "Emotions and Emotional Communication in Infants," *American Psychologist* 44, no. 2 (1989): 112.

27. Allan N. Schore, "The Effects of Early Relational Trauma on Right Brain Development, Affect Regulation, and Infant Mental Health," *Infant Mental Health Journal* 22, nos. 1–2 (2001): 201–269, 210.

28. Schore, *Affect Dysregulation and Disorders of the Self.*

29. Michael J. Meaney, "Maternal Care, Gene Expression, and the Transmission of Individual Differences in Stress Reactivity across Generations," *Annual Review of Neuroscience* 24, no. 1 (2001): 116–192.

30. Narvaez, *Neurobiology and the Development of Human Morality.*

31. M. Main and J. Solomon, "Discovery of an Insecure Disoriented Attachment Pattern: Procedures, Findings and Implications for the Classification of Behavior," in T. Berry Brazelton and Michael W. Yogman, *Affective Development in Infancy,* (Norwood, NJ: Ablex, 1986)

32. Schore, "Effects of Early Relational Trauma," 201–269, 212.

33. Krystal quoted in Gay A. Bradshaw, *Elephants on the Edge: What Animals Teach Us about Humanity* (New Haven: Yale University Press, 2009), 134.

34. Henry Krystal, "Optimizing Affect Function in the Psychoanalytic Treatment of Trauma," in Danielle Knafo, ed., *Living with Terror, Working with Trauma: A Clinician's Handbook* (Lanham, MD: Jason Aronson, 2004): 283–296.

35. John Hargrove and Howard Chua-Eoan, *Beneath the Surface: Killer Whales, SeaWorld, and the Truth beyond "Blackfish"* (New York: Macmillan, 2015); Ric O-Barry's Dolphin Project, "No Orca Dies Peacefully Say Former Trainers," March 10, 2016, https://dolphinproject.net/blog/post/op-ed-no-orca-dies-peacefully-say-former-trainers, accessed March 11, 2016.

36. Tim Ingold, "On the Social Relations of the Hunter-Gatherer Band," in Richard B. Lee and Richard Daly, eds., *The Cambridge Encyclopedia of Hunters and Gatherers* (Cambridge, Eng.: Cambridge University Press, 1999): 399–410; Darcia Narvaez, "The 99 Percent—Development and Socialization within an Evolutionary Context: Growing Up to Become 'A Good and Useful Human Being,'" in D. Fry, ed., *War, Peace and Human Nature: The Convergence of Evolutionary and Cultural Views* (New York: Oxford University Press, 2013).

37. Narvaez, "The 99 Percent," 341.

38. Narvaez, *Neurobiology.*

39. Ibid.

40. Ingold, "On the Social Relations of the Hunter-Gatherer Band"; Narvaez, "The 99 Percent."

41. Bradshaw, *Elephants on the Edge,* 134.

42. Meaney, "Maternal Care."

43. Robert A. Hinde, "Attachment: Some Conceptual and Biological Issues," in Colin Murray Parkes and J. S. Hinde, *The Place of Attachment in Human Behavior* (New York: Basic Books, 1982), 70–71.

44. Jay Belsky, "Etiology of Child Maltreatment: A Developmental-Ecological Analysis," *Psychological Bulletin* 114, no. 3 (1993): 413.

45. Jay Belsky, "War, Trauma and Children's Development: Observations from a Modern Evolutionary Perspective," *International Journal of Behavioral Development* 32, no. 4 (2008): 260–271.

46. Jeffry A. Simpson, "Attachment Theory in Modern Evolutionary Perspective," in J. Cassidy and P. R. Shaver, eds., *Handbook of Attachment: Theory, Research, and Clinical Applications* (New York: Guilford Press, 1999), 115–140.

47. Tvoparents, *Dr. Gabor Maté: Consequences of Stressed Parenting,* 2012, https://www.youtube.com/watch?v=UGmADfU5HGU, accessed July 19, 2016.

48. S. Oyama, P. E. Griffiths, and R. D. Gray, *Cycles of Contingency: Developmental Systems and Evolution* (Cambridge, MA: MIT Press, 2001); Darcia Narvaez et al., "The Evolved Development Niche: Longitudinal Effects of Caregiving Practices on Early Childhood Psychosocial Development," *Early Childhood Research Quarterly* 28, no. 4 (2013): 759–773.

49. Robin I. M. Dunbar, "The Social Brain Hypothesis and Its Implications for Social Evolution," *Annals of Human Biology* 36, no. 5 (2009): 562–572.

50. D. Narvaez et al., "Flourishing as an Aim of Child Development," in D. Narvaez et al., eds., *Contexts for Young Child Flourishing: Evolution, Family and Society* (New York: Oxford University Press, forthcoming 2016), 25.

51. Darcia Narvaez et al., "The Evolved Development Niche: Longitudinal Effects of Caregiving Practices on Early Childhood Psychosocial Development," *Early Childhood Research Quarterly* 28, no. 4 (2013): 759–773.

52. Alice Miller, *For Your Own Good: Hidden Cruelty in Child-Rearing and the Roots of Violence* (London: Macmillan, 2002).

53. Ibid; Katharina Rutschky, "Schwarze Pädagogik," in Rutschky, *Quellen zur Naturgeschichte der bürgerlichen Erziehung* (Frankfurt am Main: Ullstein, 1977).

54. F. C. Craighead, Jr., and J. J. Craighead, "Grizzly Bear Prehibernation and Denning Activities as Determined by Radiotracking," *Wildlife Monographs,* 32 (1972): 3–35.

55. Haroldson, "Grizzly Bear."

56. Robbins quoted in Richard Cockle, "Bad Year for Grizzlies with Two Fatalities in Yellowstone and 10 Other Attacks in the West," *OregonLive.com,* last updated December 28, 2011, http://www.oregonlive.com/pacific-northwest-news/index.ssf/2011/12/bad_year_for_grizzlies_with_tw.html, accessed June 27, 2016.

57. While biologists generally eschew using the term to describe the aggregate functions of wildlife over space and time, "culture" is being increasingly applied to sperm whales, elephants, and other social species. See Hal Whitehead, "Society and Culture in the Deep and Open Ocean: The Sperm Whale and Other Cetaceans," in *Animal Social Complexity: Intelligence, Culture, and Individualized Societies* (Cambridge, MA: Harvard University Press, 2005); and Gay A. Bradshaw, Allan N. Schore, Janine L. Brown, Joyce H. Poole, and Cynthia J. Moss, "Elephant Breakdown," *Nature* 433, no. 7028 (2005): 807.

58. Allan L. Egbert and Allen W. Stokes, "The Social Behavior of Brown Bears in an Alaska Salmon Stream," in M. R. Pelton, J. W. Lentfer, and G. E. Folks, eds., *Bears: Their Ecology and Management,* n.s. 40 (Morges, Switz.: IUCN, 1971), 41–56.

59. Charlie Russell, in discussion with the author, May 17, 2013.

60. Charlie Russell, in discussion with the author, July 2012.

61. Lynn L. Rogers, "Effects of Food Supply and Kinship on Social Behavior, Move-

ments, and Population Growth of Black Bears in Northeastern Minnesota," *Wildlife Monographs* (1987): 3–72.

62. Charlie Vandergaw, in discussion with the author, June 8, 2014.

63. Charlie Russell, in discussion with the author, August 23, 2013.

64. Albert Memmi, *The Colonizer and the Colonized* (New York: Routledge, 2013).

65. Maria Yellow Horse Brave Heart, "The Historical Trauma Response among Natives and Its Relationship with Substance Abuse: A Lakota Illustration," *Journal of Psychoactive Drugs* 35, no. 1 (2003): 7–13.

66. Charlie Russell, in discussion with the author, October 2014.

Chapter 3. Orcas

1. Fabienne Delfour and Denise Herzing, "Underwater Mirror Exposure to Free-Ranging Naïve Atlantic Spotted Dolphins (*Stenella frontalis*) in the Bahamas," *International Journal of Comparative Psychology* 26 (2013): 158–165; Fabienne Delfour and Ken Marten, "Mirror Image Processing in Three Marine Mammal Species: Killer Whales (*Orcinus orca*), False Killer Whales (*Pseudorca crassidens*), and California Sea Lions (*Zalophus californianus*)," *Behavioural Processes* 53, no. 3 (2001): 181–190.

2. Center for Whale Research, "About Killer Whales," http://www.whaleresearch.com, accessed August 23, 2015.

3. Justo Oxa quoted in Marisol de la Cadena, *Earth Beings: Ecologies of Practice across Andean Worlds* (Durham, N.C.: Duke University Press, 2015). Emphasis mine and de la Cadena's.

4. Marisol de la Cadena, *Earth Beings: Ecologies of Practice across Andean Worlds* (Durham, N.C.: Duke University Press, 2015).

5. Marisol de la Cadena, "About 'Mariano's Archive': Ecologies of Stories," in Lesley Green, ed., *Contested Ecologies: Dialogues in the South on Nature and Knowledge* (Cape Town: HSRC Press, 2013), 59.

6. Inge Bolin, "Chillihuani's Culture of Respect and the Circle of Courage," *Reclaiming Children and Youth* 18, no. 4 (2010): 12.

7. Ibid., 13.

8. Darcia Narvaez, *Neurobiology and the Development of Human Morality: Evolution, Culture, and Wisdom* (New York: W. W. Norton, 2014).

9. Delfour and Herzing, "Underwater Mirror."

10. Luke Rendell and Hal Whitehead, "Culture in Whales and Dolphins," *Behavioral and Brain Sciences* 24, no. 2 (2001): 309–324.

11. Hal Whitehead and Luke Rendell, *The Cultural Lives of Whales and Dolphins* (Chicago: University of Chicago Press, 2014), 11.

12. J. W. Durban and R. L. Pitman, "Antarctic Killer Whales Make Rapid, Round-

Trip Movements to Subtropical Waters: Evidence for Physiological Maintenance Migrations?" *Biology Letters* 8, no. 2 (2012): 274–277; Orca Network, "Welcome to Orca Network," http://www.orcanetwork.org, accessed June 27, 2016.

13. Durban and Pitman, "Antarctic Killer Whales," 274.

14. Orca Network, "Welcome to Orca Network."

15. Center for Whale Research, "Center for Whale Research," www.whaleresearch.com, accessed July 19, 2016.

16. Orca Network, "Offshores," March 2016, http://www.orcanetwork.org/nathist/offshores.html, accessed June 27, 2016.

17. Howard Garrett, in discussion with the author, March 2016.

18. A. D. Foote et al., "Tracking Niche Variation over Millennial Timescales in Sympatric Killer Whale Lineages," *Proceedings of the Royal Society B: Biological Sciences* 280 (2013), http://rspb.royalsocietypublishing.org/content/280/1768/20131481, accessed June 27, 2016.

19. Stuart Blackman, "Evidence for Sympatric Speciation," *The Scientist,* last modified February 9, 2006, http://www.the-scientist.com/?articles.view/articleNo/23704/title/Evidence-for-sympatric-speciation, accessed August 28, 2015.

20. Morin is quoted in Virginia Morell, "North Atlantic Killer Whales May Be Branching into Two Species," *Science Magazine Daily News,* last updated August 16, 2013, http://news.sciencemag.org/evolution/2013/08/north-atlantic-killer-whales-may-be-branching-two-species, accessed August 9, 2015.

21. Foote et al., "Tracking Niche Variation."

22. Hal Whitehead, "Cultural Selection and Genetic Diversity in Matrilineal Whales," *Science* 282, no. 5394 (1998): 1708–1711.

23. Center for Whale Research, "Southern Resident Killer Whales," http://www.whaleresearch.com/orca-population, accessed July 10, 2016.

24. Laure Frésard et al., "Epigenetics and Phenotypic Variability: Some Interesting Insights from Birds," *Genetics, Selection, Evolution* 45, no. 1 (2013), http://www.ncbi.nlm.nih.gov/pmc/articles/PMC3693910, accessed June 27, 2016.

25. De la Cadena, "About 'Mariano's Archive,'" 59.

26. Roy Wagner, "The Fractal Person," in Maurice Godelier and Marilyn Stratern, eds., *Big Men and Great Men* (Cambridge, Eng.: Cambridge University Press, 1991), 163.

27. Howard Garrett, in discussion with the author, June 2015.

28. Luciano Dalla Rosa et al., "*Orcinus orca:* A Species Complex," unpublished report of the Whale and Dolphin Conservation Society; Marilyn E. Dahlheim and Paula A. White, "Ecological Aspects of Transient Killer Whales *Orcinus orca* as Predators in Southeastern Alaska," *Wildlife Biology* 16, no. 3 (2010): 308–322; Tatiana Ivkovich et al., "The Social Organization of Resident-Type Killer Whales (*Orcinus orca*) in Avacha Gulf, Northwest Pacific, as Revealed through Association Patterns and Acous-

tic Similarity," *Mammalian Biology-Zeitschrift für Säugetierkunde* 75, no. 3 (2010): 198–210; Howard Garrett, in discussion with the author, March 2017.

29. Colin A. Simpfendorfer, Adrian B. Goodreid, and Rory B. McAuley, "Size, Sex and Geographic Variation in the Diet of the Tiger Shark, *Galeocerdo Cuvier,* from Western Australian Waters," *Environmental Biology of Fishes* 61, no. 1 (2001): 37–46; Howard Garrett, in discussion with the author, March 2016.

30. Mark McCormick, "Wild Orcas in the Pacific Northwest Are Starving to Extinction. Here's What You Can Do to Help," *One Green Planet,* last updated March 20, 2015, http://www.onegreenplanet.org/animalsandnature/wild-orcas-in-the-pacific-northwest-are-starving-to-extinction, accessed August 28, 2015.

31. John Harrison, "June Hogs," *The Northwest Power and Conservation Council,* last updated October 31, 2008, https://www.nwcouncil.org/history/JuneHogs, accessed August 28, 2015.

32. John Kenneth Baker Ford et al., *Prey Selection and Food Sharing by Fish-Eating "Resident" Killer Whales (Orcinus orca) in British Columbia* (Ottawa: Fisheries and Oceans Canada, Science, 2005).

33. Howard Garrett, July 2015, in discussion with the author.

34. "One Endangered Species Eats Another: Killer Whales and Salmon," *NOAA Fisheries,* last updated January 22, 2013, http://www.nmfs.noaa.gov/stories/2013/01/1_22_13killer_whale_chinook.html, accessed August 28, 2015; "Survival of Endangered Orcas in the Salish Sea Depends on Restoring Chinook," *Bellingham Herald,* last updated February 27, 2015, http://www.bellinghamherald.com/opinion/article22278003.html, accessed August 28, 2015.

35. Howard Garrett, "Revealing New Data Shows Killer Whales' Affinity for the Columbia River Mouth," *Save Our Wild Salmon,* http://www.wildsalmon.org/projects/the-orca-connection/revealing-new-data-shows-puget-sound-s-killer-whales-affinity-for-the-columbia-river-mouth.html, accessed August 28, 2015.

36. Ken Balcomb, in discussion with the author, June 11, 2015.

37. Robert McClure, "Are the Orcas Starving?" *Seattle Post-Intelligencer,* October 24, 2008, http://www.seattlepi.com/local/article/Are-the-orcas-starving-1289316.php, accessed November 5, 2015.

38. Emma A. Foster et al., "Social Network Correlates of Food Availability in an Endangered Population of Killer Whales, *Orcinus Orca,*" *Animal Behaviour* 83, no. 3 (2012): 731–736.

39. Howard Garrett, in discussion with the author, July 2015.

40. C. Guinet et al., "Long-Term Studies of Crozet Island Killer Whales Are Fundamental to Understanding the Economic and Demographic Consequences of Their Depredation Behaviour on the Patagonian Toothfish Fishery," *Journal of Marine Science* 72, no. 5 (2015): 1587–1597.

41. Guido di Prisco, Ennio Cocca, Sandra K. Parker, and H. William Detrich, "Tracking the Evolutionary Loss of Hemoglobin Expression by the White-Blooded Antarctic Icefishes," *Gene* 295, no. 2 (2002): 185–191.

42. "What Are Toothfish and Why Defend Them?" *Sea Shepherd,* http://www.seashepherdglobal.org/icefish/about-the-campaign/defending-toothfish.html, accessed August 28, 2015.

43. Ibid.

44. P. Tixier et al., "Interactions of Patagonian Toothfish Fisheries with Killer and Sperm Whales in the Crozet Islands Exclusive Economic Zone: An Assessment of Depredation Levels and Insights on Possible Mitigation Strategies," *CCAMLR Science* 17 (2010): 179–195.

45. Orca Aware, "Howard Garrett and the Orca Network," http://www.orcaaware.org/12/post/2012/09/howard-garrett-and-the-orca-network.html, accessed August 28, 2015; Howard Garrett, in discussion with the author, June 2015.

46. Narvaez, *Neurobiology and the Development of Human Morality,* 248.

47. Bolin, "Chillihuani's Culture," 13.

48. Narvaez, *Neurobiology and the Development of Human Morality,* 54–55.

49. Ibid., 193.

50. A. N. Schore in ibid., xvi.

51. Bolin, "Chillihuani's Culture."

52. Howard Garrett, in discussion with the author, July 2015.

53. H. Yurk et al., "Cultural Transmission within Maternal Lineages: Vocal Clans in Resident Killer Whales in Southern Alaska," *Animal Behaviour* 63, no. 6 (2002): 1103–1119.

54. Karen McComb et al., "Matriarchs as Repositories of Social Knowledge in African Elephants," *Science* 292, no. 5516 (2001): 491–494.

55. In one incident, a diver was dragged down from ocean surface then released forty seconds later, but this was considered an accident with the orca mistaking the black-suited diver for a seal. See David Kirby, "Did a Wild Orca Really Just Attack a Diver in New Zealand?," last updated February 24, 2014, http://www.takepart.com/article/2014/02/24/did-wild-orca-really-just-attack-diver-new-zealand, accessed August 4, 2015.

56. Whale and Dolphin Conservation, "The Penn Cove Orca Captures," http://us.whales.org/issues/penn-cove-orca-captures, accessed August 2, 2015.

57. Ibid.

58. Howard Garrett, in discussion with the author, July 2015.

59. Orca Network, "Proposal to Retire the Orca Lolita to Her Native Habitat in the Pacific Northwest," http://www.orcanetwork.org/Main/index.php?categories_file=Lolita, accessed November 2015.

60. Vivian Kuo, "Orca Trainer Saw Best of Keiko, Worst of Tilikum," *CNN,* last updated October 28, 2013, http://www.cnn.com/2013/10/26/world/americas/orca-trainer-tilikum-keiko/index.html, accessed August 5, 2015.

61. Ibid.

62. Ibid.

63. Christine Lozier-Duprey, *ACTUAL FOOTAGE—KILLER WHALES Mangle 1 Trainer & Nearly Kills Another ORCAS ORCA at SeaWorld,* February 28, 2014, https://www.youtube.com/watch?v=EYTIZxiIXc8, accessed June 8, 2015.

64. Paul Thompson, "Sea World Killer Whale Attack: Trainer Dawn Brancheau Was Still Vulnerable," *Daily Mail Online,* last updated February 26, 2010, http://www.dailymail.co.uk/news/article-1253881/Sea-World-killer-whale-attack-Trainer-Dawn-Brancheau-vulnerable.html, accessed August 1, 2015.

65. CNN, *SeaWorld Releases Video of 2006 Killer Whale Attack,* last updated July 25, 2012, https://www.youtube.com/watch?v=5B_poyjBqYE, accessed August 5, 2015.

66. Ed Pilkington, "Killer Whale Tilikum to Be Spared after Drowning Trainer by Ponytail," *The Guardian,* last updated February 25, 2010, http://www.theguardian.com/world/2010/feb/25/killer-whale-tilikum-drowned-trainer-hair, accessed August 4, 2015.

67. See letter sent by an observer of Dawn Brancheau's death to the Orange County Sheriff's office, quoted in Tim Zimmermann, "Diary of a Killer Whale: What Motivated Tilikum's Attack on Dawn Brancheau?," July 8, 2010, http://timzimmermann.com/2010/07/08/diary-of-a-killer-whale-what-motivated-tilikums-attack-on-dawn-brancheau, accessed August 9, 2015.

68. Howard Garrett, in discussion with the author, July 2015.

69. Fred Buyle, in discussion with the author, June 2015.

70. "Killers of Eden," http://www.killersofeden.com, accessed August 4, 2015.

71. Howard Garrett, in discussion with the author, June 2015.

72. Lucien Lévy-Bruhl, *How Natives Think* (Princeton, N.J.: Princeton University Press, 1985).

73. Namu Griff, *Orca Semen Collecting,* April 5, 2007, https://www.youtube.com/watch?v=TIU2-m_Vc7U, accessed July 9, 2015.

74. Thomas J. Csordas, "Somatic Modes of Attention," *Cultural Anthropology* 8, no. 2 (1993): 135–156.

75. Debra Niehoff, *The Biology of Violence: How Understanding the Brain, Behavior, and Environment Can Break the Vicious Circle of Aggression* (New York: Free Press, 1999), 185.

76. The Orca Project, "Exclusive Interview #3: Former SeaWorld Trainer Samantha Berg and the Perils of Orca Captivity (part 2)," March 14, 2011, https://theorcaproject.wordpress.com/2011/03/14/exclusive-interview-former-seaworld-trainer-samantha-berg-2, accessed July 12, 2015.

77. Howard Garrett, in discussion with the author, June 2015.

78. Emily Dickinson, "I Felt a Funeral, In My Brain," http://www.poetryfoundation.org/poems-and-poets/poems/detail/45706, accessed July 18, 2016.

79. NOAA, "Captive Killer Whale Included in Endangered Listing," February 4, 2015. http://www.noaanews.noaa.gov/stories2015/20150204-captive-killer-whale-included-in-endangered-listing.html, accessed June 27, 2016.

80. Teresa Demerast, "Why Is the Miami Seaquarium Lying about Keiko?" *Huffington Post,* February 12, 2015, http://www.huffingtonpost.com/theresa-demarest/why-is-the-miami-seaquari_b_6648938.html, accessed March 12, 2016.

81. Howard Garrett, in discussion with the author, July 2016.

82. Ken Balcomb quoted in Paul Watson, "The Life of Tilikum: Fear and Loathing, Deprivation and Death in the Aquatic Asylum," *The Dodo,* March 18, 2016, http://www.onegreenplanet.org/animalsandnature/the-sad-life-of-tilikum/?utm_source=Green+Monster+Mailing+List&utm_campaign=661db10e0a-NEWSLETTER_EMAIL_CAMPAIGN&utm_medium=email&utm_term=0_bbf62ddf34–661db10e0a-105997921, accessed March 18, 2016.

83. Tim Zimmerman, "Tilikum Is Dying," *National Geographic,* March 10, 2016, http://news.nationalgeographic.com/2016/03/160310-tilikum-killer-whale-orca-death-seaworld-sick-dying, accessed March 10. 2016; "Tilikum the Killer Whale. . ." *Washington Post,* March 9, 2016, https://www.washingtonpost.com/news/morning-mix/wp/2016/03/09/tilikum-the-killer-whale-at-the-center-of-blackfish-may-be-in-his-final-days-after-decades-of-controversy-and-three-deaths, accessed July 10, 2016.

84. Joel Manby, "SeaWorld CEO: We're Ending Our Orca Breeding Program. Here's Why," *Los Angeles Times,* March 19, 2016, http://www.latimes.com/opinion/op-ed/la-oe-0317-manby-sea-world-orca-breeding-20160317-story.html, accessed July 19, 2016.

Chapter 4. Crocodiles

1. Walter Auffenberg, *The Behavioral Ecology of the Komodo Monitor* (Gainesville: University Press of Florida, 1981), 133.

2. Erich Jarvis, in discussion with the author, May 2013.

3. E. D. Jarvis et al., "Avian Brains and a New Understanding of Vertebrate Brain Evolution," *Nature Reviews Neuroscience* 6, no. 2 (2005): 151–159, http://www.ncbi.nlm.nih.gov/pmc/articles/PMC2507884, accessed June 27, 2016.

4. "A rose is a rose is a rose is a rose," from the poem "Sacred Emily" in Gertrude Stein, *Selected Writings of Gertrude Stein* (New York: Vintage Books, 1990); Babette S. Hellemans, *Rethinking Abelard: A Collection of Critical Essays* (Leiden, Neth.: Brill, 2014); John Marenbon, *The Philosophy of Peter Abelard* (Cambridge, Eng.: Cambridge University Press, 1997).

5. Paul D MacLean, *The Triune Brain in Evolution: Role in Paleocerebral Functions* (New York: Plenum Press, 1990).

6. Bjorn Merker, "Consciousness without a Cerebral Cortex: A Challenge for Neuroscience and Medicine," *Behavioral and Brain Sciences* 30, no. 1 (2007): 63.

7. Jarvis, "Avian Brains," 151.

8. Carl Sagan, *Cosmos,* vol. 1 (Barcelona: Edicions de la Universitat Barcelona, 2006).

9. Tadashi Nomura et al., "Reptiles: A New Model for Brain Evo-Devo Research," *Journal of Experimental Zoology, Part B: Molecular and Developmental Evolution* 320, no. 2 (March 2013): 57–73.

10. Auffenberg, *Behavioral Ecology,* 138.

11. One Saltwater observed off Australia's famous Broome Cable Beach was actually doing what can only be called surfing. See "*Great White Shark . . . No ! Hungry 13FT Monster CROCODILE Surfing and Closes a Beach in Australia, Patryn WorldLatest-New,* accessed June 28, 2015.

12. Steven G. Platt and John B. Thorbjarnarson, "Nesting Ecology of the American Crocodile in the Coastal Zone of Belize," *Copeia* 3 (2000): 869–873; J. B. Thorbjarnarson, "Ecology of the American Crocodile, *Crocodylus acutus,*" in *Proceedings of the 7th Working Meeting of the Crocodile Specialist Group,* ed. Crocodile Specialist Group (Gland, Switz.: International Union for Conservation of Nature, 1986).

13. Fred Buyle, in discussion with the author, June 2015.

14. Auffenberg, *Behavioral Ecology,* 138.

15. Matthew H. Shirley, Kent A. Vliet, Amanda N. Carr, and James D. Austin, "Rigorous Approaches to Species Delimitation Have Significant Implications for African Crocodilian Systematics and Conservation," *Proceedings of the Royal Society of London B: Biological Sciences* 281, no. 1776 (2014), DOI: 10.1098/rspb. 2013.2483; Evon Hekkala et al., "An Ancient Icon Reveals New Mysteries: Mummy DNA Resurrects a Cryptic Species within the Nile Crocodile," *Molecular Ecology* 20, no. 20 (2011): 4199–4215.

16. Y. Milián-García et al., "Evolutionary History of Cuban Crocodiles *Crocodylus rhombifer* and *Crocodylus acutus* Inferred from Multilocus Markers," *Journal of Experimental Zoology, Part A: Ecological Genetics and Physiology* 315, no. 6 (2011): 358–375.

17. S. G Platt et al., "Frugivory and Seed Dispersal by Crocodilians: An Overlooked Form of Saurochory?," *Journal of Zoology* 291, no. 2 (2013): 87–99.

18. "Crocodile Confession: Meat-Eating Predators Consume Fruit, Study Says," *Wildlife Conservation Society,* last modified August 21, 2013, http://www.wcs.org/press/press-releases/crocodile-confession.aspx, accessed August 2, 2015.

19. Adam E. Rosenblatt, Scott Zona, Michael R. Heithaus, and Frank J. Mazzotti, "Are Seeds Consumed by Crocodilians Viable? A Test of the Crocodilian Saurochory Hypothesis," *Southeastern Naturalist* 13, no. 3 (2014): N26–N29.

20. "*Crocodylus moreletii,*" *Crocodilians,* http://crocodilian.com/cnhc/csp_cmor.htm, accessed August 29, 2015.

21. Steven G. Platt et al., "Consumption of Large Mammals by *Crocodylus moreletii:* Field Observations of Necrophagy and Interspecific Kleptoparasitism," *Southwestern Naturalist* 52, no. 2 (2007): 310–317.

22. Ibid.

23. Kevin Hansen and Robert Redford, *Cougar: The American Lion* (Flagstaff, AZ: Northland Press, 1992); Mel Sunquist, "Cougar: The American Lion," *Journal of Mammalogy* 75, no. 1 (1994): 233–234.

24. Earl Showerman, in discussion with the author, June 24, 2015.

25. Platt et al., "Consumption."

26. Daniel Goleman, *Emotional Intelligence* (New York: Bantam Books, 2006).

27. George Loewenstein and Jennifer S. Lerner, "The Role of Affect in Decision Making," in Richard J. Davidson, Klaus R. Scherer, and H. Hill Goldsmith, eds., *Handbook of Affective Sciences,* (New York: Oxford University Press, 2003).

28. Daniel Goleman, *Emotional Intelligence* (New York: Bantam Books, 2006).

29. Val Plumwood, *The Eye of the Crocodile,* ed. Lorraine Shannon (Canberra: ANU E Press, 2013).

30. Sarda Sahney and Michael J. Benton, "Recovery from the Most Profound Mass Extinction of All Time," *Proceedings of the Royal Society of London B: Biological Sciences* 275, no. 1636 (2008): 759–765; Zhong-Qiang Chen and Michael J. Benton, "The Timing and Pattern of Biotic Recovery Following the End-Permian Mass Extinction," *Nature Geoscience* 5, no. 6 (2012): 375–383.

31. Colleen G. Farmer, "The Evolution of Unidirectional Pulmonary Airflow," *Physiology* 30, no. 4 (2015): 260–272.

32. J. Michael Parrish, "The Origin of Crocodilian Locomotion," *Paleobiology* (1987): 396–314; Adam P. Summers, "Evolution: Warm-Hearted Crocs," *Nature* 434, no. 7035 (2005): 833–834.

33. Albert Bandura, "Regulation of Cognitive Processes through Perceived Self-Efficacy," *Developmental Psychology* 25, no. 5 (1989): 729.

34. U. de V. Pienaar, "Predator-Prey Relationships among the Larger Mammals of the Kruger National Park," *Koedoe* 12 (1969): 108–176.

35. V. Dinets, J. C. Brueggen, and J. D. Brueggen, "Crocodilians Use Tools for Hunting," *Ethology Ecology & Evolution* 27, no. 1 (2015): 74–78.

36. "Learn about the Everglades in an Up-Close and Personal Way," *Everglades Outpost,* http://www.evergladesoutpost.org, accessed August 9, 2015.

37. Bob Freer, in discussion with the author, May 2015.

38. Matthew L. Brien et al., "Intra- and Interspecific Agonistic Behaviour in Hatchling Australian Freshwater Crocodiles (*Crocodylus johnstoni*) and Saltwater Crocodiles

(*Crocodylus porosus*)," *Australian Journal of Zoology* 61, no. 3 (2013): 196–205, http://www.publish.csiro.au/paper/ZO13035.htm, accessed June 27, 2016; Matthew L. Brien et al., "Born to Be Bad: Agonistic Behaviour in Hatchling Saltwater Crocodiles (*Crocodylus porosus*)," *Behaviour* 150, no. 7 (2013): 737–762.

39. P. Charruau and Y. Henaut, "Nest Attendance and Hatchling Care in Wild American Crocodiles (*Crocodylus acutus*) in Quintana Roo, Mexico," *Animal Biology* 62, no. 1 (2012): 29–51; V. Dinets, "Play Behavior in Crocodilians," *Animal Behavior and Cognition* 2, no. 1 (2015): 49–55.

40. Gordon M. Burghardt, *The Genesis of Animal Play: Testing the Limits* (Cambridge, MA: MIT Press, 2005), 98; James B. Murphy, *Komodo Dragons: Biology and Conservation* (Washington, DC: Smithsonian Institution Press, 2002).

41. "Chito and Pocho," *NPR.org*, last updated February 28, 2014, http://www.npr.org/2014/02/28/283934611/chito-and-pocho, accessed August 1, 2015.

42. Dinets, "Play Behavior," 54.

43. Matt Levin, "Costa Rica Says Goodbye to Famous Croc Pocho," *Tico Times*, last updated October 17, 2011, http://www.ticotimes.net/2011/10/17/costa-rica-says-goodbye-to-famous-croc-pocho, accessed August 29, 2015.

44. Ibid.

45. A. Wilkinson et al., "Social Learning in a Non-Social Reptile (*Geochelone carbonaria*)," *Biology Letters* 6, no. 5 (2010): 614–616.

46. K. M. Davis and G. M. Burghardt, "Turtles (*Pseudemys nelsoni*) Learn about Visual Cues Indicating Food from Experienced Turtles," *Journal of Comparative Psychology* 125, no. 4 (2011): 404–410; J. Sean Doody, Gordon M. Burghardt, and Vladimir Dinets, "Breaking the Social–Non-Social Dichotomy: A Role for Reptiles in Vertebrate Social Behavior Research?," *Ethology* 119, no. 2 (2013): 95–103; Gordon M. Burghardt, "Learning Processes in Reptiles," *Biology of the Reptilia* 7, no. 7 (1977): 555–681.

47. David J. Varricchio et al., "Mud-Trapped Herd Captures Evidence of Distinctive Dinosaur Sociality," *Acta Palaeontologica Polonica* 53, no. 4 (2008): 567–578.

48. Vladimir Dinets, "Apparent Coordination and Collaboration in Cooperatively Hunting Crocodilians," *Ethology Ecology and Evolution* 27, no. 2 (2015): 244–250.

49. "Crocodiles, Alligators Hunt in Groups, Scientist Says," *SciNews*, October 14, 2014, http://www.sci-news.com/biology/science-crocodiles-alligators-hunt-groups-02203.html, accessed June 25, 2016.

50. Lawrence Henriques, in discussion with the author, June 2015.

51. Z. Campos et al., "Parental Care in the Dwarf Caiman, *Paleosuchus palpebrosus*, Cuvier, 1807 (Reptilia: Crocodilia: Alligatoridae)," *Journal of Natural History* 46, nos. 47–48 (2012): 2979–2984.

52. Howard R. Hunt, "Nest Excavation and Neonate Transport in Wild *Alligator mississippiensis*," *Journal of Herpetology* (1987): 348–350; Doody et al., "Breaking the Social–Non-Social Dichotomy," 4.

53. P. C. Watts et al., "Parthenogenesis in Komodo Dragons.," *Nature* 444, no. 7122 (2006): 1021–1022.

54. Bob Freer, in discussion with the author, June 22, 2015.

55. David McFadden, "Crocodiles Disappearing as Dinner in Jamaica," *Yahoo News,* last update October 6, 2013, http://news.yahoo.com/crocodiles-disappearing-dinner-jamaica-150218364.html, accessed August 5, 2015.

56. Lawrence Henriques, in discussion with the author, June 2015.

57. Jessica Bell Rizzolo and G. A. Bradshaw, "Sustainability Inside and Out: Including Psychological and Social Criteria for IUCN Asian Wildlife Assessments," paper presented at Conservation Asia 2016, Singapore; Jessica Bell Rizzolo and G. A. Bradshaw, "Sustaining Wildlife Minds and Relationships: Why Psychological and Social Criteria Are Essential for IUCN RedList Criteria," *Conservation Biology* (in press).

58. Lawrence Henriques, in discussion with the author, June 2015.

59. Rob Denkhaus, in discussion with the author, May 2015.

60. Lawrence Henriques, in discussion with the author, July 2015.

61. Ibid.

62. Ibid.

63. Monika Böhm et al., "The Conservation Status of the World's Reptiles," *Biological Conservation* 157 (2013): 372–385.

64. Anslem Da Silva and Janaki Lenin, "Mugger Crocodile *Crocodylus palustris,*" *Crocodiles. Status Survey and Conservation Action Plan. Crocodile Specialist Group, Darwin, Australia* (2010): 94–98; C. N. Kumar and D. Gopal, "Indian Gharial (*Gavialis gangeticus*) on the Verge of Extinction," *Current Science* 94, no. 12 (2008): 1549.

65. Mwelwa C. Musambachime, "The Fate of the Nile Crocodile in African Waterways," *African Affairs* (1987): 197–207.

66. Lewis Carroll, *Alice in Wonderland* (New York: Random House, 1955).

67. A. D. Graham and Peter H. Beard, *Eyelids of Morning: The Mingled Destinies of Crocodiles and Men; Being a Description of the Origins, History, and Prospects of Lake Rudolf, Its Peoples, Deserts, Rivers, Mountains, and Weather* (Greenwich, CT: New York Graphic Society, 1973), 12.

68. Ibid., 11; Mwelwa C. Musambachime, "The Fate of the Nile Crocodile in African Waterways," *African Affairs* (1987): 197–207.

69. Fred Buyle, in discussion with the author, June 2015.

70. Brooke is quoted in Lucien Lévy-Bruhl, *How Natives Think* (Princeton, NJ: Princeton University Press, 1985), 94.

71. Sidi Munan, "See You Later Alligator—in a While, Crocodile," *BorneoPost Online,* last modified August 25, 2013, http://www.theborneopost.com/2013/08/25/see-you-later-alligator-in-a-while-crocodile, accessed August 1, 2015.

72. Mwelwa C. Musambachime, "The Fate of the Nile Crocodile in African Waterways." *African Affairs* (1987): 197–207.

73. Ibid., 199.

74. See CrocBITE, Worldwide Crocodilian Database, http://www.crocodile-attack.info/about/frequently-asked-questions, accessed March 2016; Crocodile Specialist Group, "Crocodile Attacks," http://www.iucncsg.org, accessed August 29, 2015.

75. A study of the Philippine Agta peoples and pythons discusses how the small build of these indigenous people makes them viable prey, well within suitable dimensions for python. But the number of confirmed attacks is very low. See Thomas N. Headland and Harry W. Greene, "Hunter-Gatherers and Other Primates as Prey, Predators, and Competitors of Snakes," *Proceedings of the National Academy of Sciences* 108, no. 52 (2011): 20865–20866.

76. Kevin M. Wallace and Alison J. Leslie, "Diet of the Nile Crocodile (*Crocodylus niloticus*) in the Okavango Delta, Botswana," *Journal of Herpetology* 42, no. 2 (2008): 361–368.

77. Brien quoted in Oliver Milman, "Saltwater Crocodile Named World's Most Aggressive," *The Guardian,* December 12, 2013, http://www.theguardian.com/environment/2013/dec/12/saltwater-crocodile-named-worlds-most-aggressive, accessed August 3, 2015.

78. Sara Nelson, "Disney Resort Attack: 4 Alligators Killed But No Sign of Missing Toddler," *Huffington Post,* June 15, 2016, http://www.huffingtonpost.co.uk/entry/disney-resort-attack-4-alligators-killed-but-no-sign-of-missing-toddler_uk_576162c4e40681487dc6428, accessed July 18, 2016; Selina Sykes, "Four Alligators Killed after Boy Dies in Lake at Orlando Disney World," *Express,* June 8, 2016, http://www.express.co.uk/news/world/681670/alligators-Orlando-Disney-World-killed-toddler, accessed July 18, 2016.

79. P. Christiansen and S. Wroe, "Bite Forces and Evolutionary Adaptations to Feeding Ecology in Carnivores," *Ecology* 88, no. 2 (2007): 347–358.

80. Bryan G. Fry et al., "A Central Role for Venom in Predation by *Varanus komodoensis* (Komodo Dragon) and the Extinct Giant *Varanus (Megalania) priscus,*" *Proceedings of the National Academy of Sciences* 106, no. 22 (2009): 8969–8974.

81. Lawrence Henriques, in discussion with the author, June 25, 2015.

82. Bob Freer, in discussion with the author, May 2015; Ed Mazza, "First-Ever Crocodile Attack on Humans in the U.S.," *Huffington Post,* last updated August 27, 2014, http://www.huffingtonpost.com/2014/08/27/crocodile-attack-florida_n_5720350.html, accessed August 29, 2015.

83. Crocodile Specialist Group, "Crocodile Attacks."

84. Crocodile Specialist Group, Home Page, http://www.iucncsg.org, accessed August 29, 2015.

85. Lawrence Henriques, in discussion with the author, June 25, 2015.

86. Ibid.

87. Val Plumwood, "Being Prey," n.d., http://www.aislingmagazine.com/aislingmagazine/articles/TAM30/ValPlumwood.html, accessed August 6, 2015.

88. Scott D. Whiting and Andrea U. Whiting, "Predation by the Saltwater Crocodile (*Crocodylus porosus*) on Sea Turtle Adults, Eggs, and Hatchlings," *Chelonian Conservation and Biology* 10, no. 2 (2011): 198–205.

89. Plumwood, *Eye of the Crocodile.*

90. Ibid.

91. Ibid.

92. Lawrence Henriques, in discussion with the author, July 2015.

Chapter 5. Rattlesnakes

1. "Online Etymology Dictionary," http://www.etymonline.com/index.php?allowed_in_frame=0&search=estivate&searchmode=none, accessed August 29, 2015.

2. In 2009, the present Latin classification replaced the older *Spermophilus beecheyi.*

3. James C. Gillingham and Randy E. Baker, "Evidence for Scavenging Behavior in the Western Diamondback Rattlesnake (*Crotalus atrox*)," *Zeitschrift für Tierpsychologie* 55, no. 3 (1981): 217–227.

4. A. L. Campbell et al., "Biological Infrared Imaging and Sensing," *Micron* 33, no. 2 (2002): 211–225.

5. Elena O. Gracheva et al., "Ganglion-Specific Splicing of TRPV1 Underlies Infrared Sensation in Vampire Bats," *Nature* 476, no. 7358 (2011): 88–91.

6. M. Ros, "Die Lippengruben der Pythonen als Temperatureorgane," *Jenaische Zeitschrift für Medizin und Naturwissen* 70 (1935): 1–32.

7. Jesse M. Meik et al., "Limitations of Climatic Data for Inferring Species Boundaries: Insights from Speckled Rattlesnakes," *PLoS ONE* 10, no. 6 (2015): e0131435, http://journals.plos.org/plosone/article?id=10.1371/journal.pone.0131435, accessed June 27, 2016.

8. W. B. Montgomery et al., "Crotalus atrox × Crotalus Horridus (Western Diamond-backed Rattlesnake × Timber Rattlesnake). NATURAL HYBRID," *Herpetological Review* 44, vol. 4 (2013): 4.

9. G. W. Schuett et al., "Social Behavior of Rattlesnakes: A Shifting Paradigm," in Gordon W. Schuett et al., *Rattlesnakes of Arizona* (Rodeo, NM: Eco Publishing, in press).

10. Harry Greene, in discussion with the author, July 6, 2015.

11. M. A. Barbour and R. W. Clark, "Ground Squirrel Tail-Flag Displays Alter Both Predatory Strike and Ambush Site Selection Behaviours of Rattlesnakes," *Pro-*

ceedings of the Royal Society of London B, Biological Sciences, 279, no. 1743 (2012): 3827–3833.

12. Harry Greene, in discussion with the author, July 6, 2015.

13. Richard C. Goris, "Infrared Organs of Snakes: An Integral Part of Vision," *Journal of Herpetology* 45, no. 1 (2011): 2–14.

14. Campbell et al., "Biological Infrared Imaging and Sensing."

15. Goris, "Infrared Organs," 11.

16. Ibid.

17. M. P. Rowe and D. H. Owings, "Probing, Assessment, and Management during Interactions between Ground Squirrels and Rattlesnakes," *Ethology,* 86, no. 3 (1990): 237–249.

18. John C Perez, Sathit Pichyangkul, and Vivian E Garcia, "The Resistance of Three Species of Warm-Blooded Animals to Western Diamondback Rattlesnake (*Crotalus atrox*) Venom," *Toxicon* 17, no. 6 (1979): 601–607.

19. This stunning video depicts an Arizona black rattlesnake hunting as a ground squirrel comes up and kicks sand at him. A western diamondback also makes the scene in an intriguing inter-species rattlesnake encounter. See "Interspecies Interactions Aplenty!" *SocialSnakes,* last updated August 20, 2012, http://blog.socialsnakes.org/interspecies-interactions-aplenty, accessed August 30, 2015.

20. A. S. Rundus et al., "Ground Squirrels Use an Infrared Signal to Deter Rattlesnake Predation," *Proceedings of the National Academy of Sciences* 104, no. 36 (2007): 14372–14376.

21. Aaron Rundus, in discussion with the author, March 2016.

22. Ibid.

23. A. S. Rundus et al., "Ground Squirrels."

24. Aaron Rundus, in discussion with the author, March 2016.

25. Ronald R. Swaisgood, Matthew P. Rowe, and Donald H. Owings, "Antipredator Responses of California Ground Squirrels to Rattlesnakes and Rattling Sounds: The Roles of Sex, Reproductive Parity, and Offspring Age in Assessment and Decision-Making Rules," *Behavioral Ecology and Sociobiology* 55, no. 1 (2003): 22–31.

26. Jérémie Teyssier et al., "Photonic Crystals Cause Active Colour Change in Chameleons," *Nature Communications* 6 (2015): 63–68.

27. "The better part of Valour, is Discretion; in the which better part, I haue saued my life" (spelling and punctuation from the *First Folio,* Shakespeare, *Henry V,* act 5, scene 3, lines 3085–3086).

28. Virginia Duncan quoted in Laurence M. Klauber, *Rattlesnakes* (Berkeley: University of California Press, 1956).

29. Ibid., 121.

30. Ibid., 179.

31. M. Schneemann et al., "Life-Threatening Envenoming by the Saharan Horned Viper (*Cerastes cerastes*) Causing Micro-Angiopathi, Coagulopathy and Acute Renal Failure: Clinical Cases and Review," *QJM: Monthly Journal of the Association of Physicians* 97, no. 11 (2004): 717–727.

32. Quoted in Goris, "Infrared Organs," 2.

33. J. Sean Doody, Gordon M. Burghardt, and Vladimir Dinets, "Breaking the Social–Non-Social Dichotomy: A Role for Reptiles in Vertebrate Social Behavior Research?," *Ethology* 119, no. 2 (2013): 96; B. B. Bowers and G. M. Burhardt, "The Scientist and the Snake: Relationships with Reptiles," in Hank Davis and Dianne A. Balfour, eds., *The Inevitable Bond: Examining Scientist-Animal Interactions* (Cambridge, Eng.: Cambridge University Press, 1992).

34. Schuett et al., *Rattlesnakes of Arizona.*

35. Melissa Amarello, in discussion with the author, June 2015.

36. John Roach, "Young Americans Geographically Illiterate, Survey Suggests," last updated May 2, 2006, http://news.nationalgeographic.com/news/2006/05/0502_060502_geography.html, accessed August 30, 2015.

37. "Can Snakes Unhinge Their Jaws? Harry Greene Explains . . . ," http://www.youtube.com/watch?v=Mm9h6KE-ZOk, accessed July 5, 2015.

38. Schuett et al., *Rattlesnakes of Arizona.*

39. Steven J. Beaupre, "Sexual Size Dimorphism in the Western Diamondback Rattlesnake (*Crotalus atrox*): Integrating Natural History, Behaviour, and Biology," *Sonoran Herpetologist* 15, no. 4 (2002); Roger A. Repp, "Herping Arizona," *Sonoran Herpetologist* 8, no. 11 (1995): 112; Roger Repp, in discussion with the author, March 2016. "The term 'stacking' was first coined in the early 1990s in Fort McDowell atrox studies conducted by Jack O'Liele and Steve Beaupre."

40. Roger Repp, in discussion with the author, July 2015.

41. Klauber, *Rattlesnakes,* 736.

42. Doody et al., "Breaking the Social—Non-Social Dichotomy."

43. Rulon W. Clark, "Kin Recognition in Rattlesnakes," *Proceedings of the Royal Society of London B: Biological Sciences* 271, suppl. 4 (2004): S243.

44. R. W. Clark et al., "Cryptic Sociality in Rattlesnakes (*Crotalus horridus*) Detected by Kinship Analysis," *Biology Letters* 8, no. 4 (2012): 523–525, http://rsbl.royalsocietypublishing.org/content/early/2012/02/13/rsbl.2011.1217, accessed June 27, 2016.

45. Charles F. Smith and Gordon W. Schuett, "Putative Pair-Bonding in *Agkistrodon contortrix* (Copperhead)," *Northeastern Naturalist* 22, no. 1 (2015): N1–N5.

46. John Scott, *Social Network Analysis* (Thousand Oaks, CA: Sage, 2012).

47. "A Rattlesnake Helper?," *SocialSnakes,* http://blog.socialsnakes.org/a-rattlesnake-helper, accessed August 30, 2015.

48. Melissa Amarello, in discussion with the author, July 2015.

49. Marco Capocasa et al., "Linguistic, Geographic and Genetic Isolation: A Collaborative Study of Italian Populations," *Journal of Anthropological Sciences* 92 (2013): 201–231.

50. Roger Repp, in discussion with the author, July 10, 2015.

51. Elizabeth Barrett Browning, "Grief," *The Poetry Foundation,* http://www.poetryfoundation.org/poem/177112, accessed August 30, 2015.

52. Darcia Narvaez, *Neurobiology and the Development of Human Morality: Evolution, Culture, and Wisdom* (New York: W. W. Norton, 2014), xxxvii.

53. Roger Repp, in discussion with the author, July 11, 2015.

54. Klauber, *Rattlesnakes,* 703.

55. Roger Repp, in discussion with the author, July 11, 2015.

56. Repp quoted in Schuett et al., "Social Behavior of Rattlesnakes."

57. Ibid.

58. Charles Darwin, "The Descent of Man. 1871," in Darwin, *The Origin of Species and the Descent of Man* (London: J. Murray, 1871).

59. Edward L. Thorndike, "Intelligence and Its Uses." *Harper's Magazine* (1920); I. M. Dunbar, "The Social Brain Hypothesis," *Foundations in Social Neuroscience* 5, no. 71 (2002): 69; Greg J. Norman et al., "Social Neuroscience and the Modern Synthesis of Social and Biological Levels of Analysis," in *Handbook of Neurosociology* (Houton, Neth.: Springer Netherlands, 2013), 67–81.

60. Robin I. M. Dunbar, "The Social Brain Hypothesis and Its Implications for Social Evolution," *Annals of Human Biology* 36, no. 5 (2009): 562–572.

61. Ibid., 562.

62. Erich Jarvis, in discussion with the author, March 2016.

63. Now it seems that snake ancestors were not aquatic but terrestrial with grasping legs and arms. See Anastasia Christakou, "Four-Legged Fossil Snake Is a World First," *Nature News,* July 23, 2015, http://www.nature.com/news/four-legged-fossil-snake-is-a-world-first-1.18050, accessed July 24, 2015.

64. Jack Weir, "The Sweetwater Rattlesnake Round-Up: A Case Study in Environmental Ethics," *Conservation Biology* 6, no. 1 (1992): 116–127; Bruce D. Means, "Effects of Rattlesnake Roundups on the Eastern Diamondback Rattlesnake (*Crotalus adamanteus*)," *Herpetological Conservation and Biology* 4, no. 2 (2009): 132–141.

65. Ibid.

66. Janet Van Vleet, "Teenage Scholarship Pageant Contestants Go Above and Beyond for College Money," *Abilene Reporter-News,* last updated February 20, 2015, http://www.reporternews.com/news/teenage-scholarship-pageant-contestants-go-above-and-beyond-for-college-money_74916791, accessed August 30, 2015.

67. Wendy Townsend, "'Rattlesnake Roundup' Teaches Cruelty Is Fun," *CNN,* http://www.cnn.com/2014/04/09/opinion/townsend-rattlesnake-roundup/index.html, accessed August 30, 2015.

68. Melissa Amarello, in discussion with the author, August 5, 2015.

69. Harry Greene, in discussion with the author, July 24, 2015.

70. Harry Greene, in discussion with the author, July 6, 2015.

71. Ibid.

72. Melissa Amarello, in discussion with the author, August 5, 2015.

73. Vicky Meretsky et al., "Quantity versus Quality in California Condor Reintroduction: Reply to Beres and Starfield," *Conservation Biology* 15 (2001): 1449–1451.

74. Brian K. Sullivan, Erika M. Nowak, and Matthew A. Kwiatkowski, "Problems with Mitigation Translocation of Herpetofauna," *Conservation Biology* 29, no. 1 (2015): 15.

75. Val Plumwood, "Being Prey," http://www.aislingmagazine.com/aislingmagazine/articles/TAM30/ValPlumwood.html, accessed August 6, 2015.

76. G. A. Bradshaw, *Elephants on the Edge: What Animals Teach Us about Humanity* (New Haven: Yale University Press, 2009).

77. Schuett et al., "Social Behavior of Rattlesnakes."

78. James Blachowicz, "There Is No Scientific Method," *New York Times,* July 4, 2016, http://www.nytimes.com/2016/07/04/opinion/there-is-no-scientific-method.html?action=click&contentCollection=opinion®ion=rank&module=package&version=highlights&contentPlacement=1&pgtype=collection&_r=, accessed July 18, 2016.

79. "Trans-Species Psychology," *Wikipedia,* January 6, 2015, https://en.wikipedia.org/w/index.php?title=Trans-species_psychology&oldid=641240173, accessed July 19, 2016.

80. David B. Prival et al., *A Comparative Study of Hunted vs. Unhunted Populations of the Twin-Spotted Rattlesnake: Final Report* (Phoenix: Arizona Game and Fish Dept., 1999), 43.

81. Yerkes National Primate Research Center, "About," http://www.yerkes.emory.edu/about, accessed September 1, 2015.

82. From an interview with Michael Lind—see Chris Lehmann, "Seeing Flowers in a New Way, through Loren Eiseley," *NPR.org,* last updated August 17, 2006, http://www.npr.org/templates/story/story.php?storyId=5634196, accessed August 30, 2015.

83. Bacon quoted in Carolyn Merchant, "The Scientific Revolution and the Death of Nature," *Isis* 97, no. 3 (2006): 518.

84. "Biography of Nikolaas Tinbergen," in Jan E. Lindsten, ed., *Nobel Lectures in Physiology or Medicine: 1971–1980* (River Edge, NJ: World Scientific, 1992), 111.

85. Nikolaas Tinbergen, "Ethology and Stress Diseases," in Jan E. Lindsten, ed., *Nobel Lectures in Physiology or Medicine: 1971–1980* (River Edge, NJ: World Scientific, 1992), 113.

86. Ibid., 128, note 1.

87. Melissa Amarello, in discussion with the author, July 2015.

88. Melissa Amarello, in discussion with the author, August 2015.

89. Ibid.

90. Ibid.

91. Ibid.

92. D. H. Lawrence, "Snake," http://homepages.wmich.edu/~cooneys/poems/dhl.snake.html, accessed July 18, 2016.

Chapter 6. Pumas

1. A. Rodrigues *Mountain Lions Killed in the U.S., 1992–2004* (Sacramento: The Mountain Lion Foundation, 2015).

2. H. B. Ernest et al., "Fractured Genetic Connectivity Threatens a Southern California Puma (*Puma concolor*) Population," *PloS One* 9, no. 10 (2014), http://journals.plos.org/plosone/article?id=10.1371/journal.pone.0107985, accessed June 27, 2016.

3. Kevin Hansen and Robert Redford, *Cougar: The American Lion* (Flagstaff, AZ: Northland Press, 1992), 57; Karen Anspacher-Meyer, *Lords of Nature: Life in a Land of Great Predators,* 2009 film documentary, http://www.imdb.com/title/tt1620879/?ref_=fn_al_tt_5, accessed June 27, 2016.

4. P. Beier, S. P. D. Riley, and Stanley D. Gehrt, "Mountain Lions (*Puma concolor*)" in Stanley D. Gehrt, Seth P. D. Riley, and Bryan L. Cypher, eds., *Urban Carnivores: Ecology, Conflict, and Conservation* (Baltimore: Johns Hopkins University Press, 2010), 141–156.

5. "Lions in the Santa Monica Mountains?," *U.S. National Park Service,* http://www.nps.gov/samo/learn/nature/pumapage.htm, accessed August 30, 2015; Harley G. Shaw, *Soul among Lions: The Cougar as Peaceful Adversary* (Tucson: University of Arizona Press, 2000), 41; M. M. Grigione et al., "Ecological and Allometric Determinants of Home-Range Size for Mountain Lions (*Puma concolor*)," *Animal Conservation* 5, no. 4 (2002): 317–324.

6. Shaw, *Soul among Lions,* 41.

7. Ibid., 43.

8. Ibid., 40.

9. Frank C. Hibben, *Hunting American Lions* (Silver City, NM: High-Lonesome Books, 1995).

10. Jay Bruce, *LionHunter,* Mountain Lion Foundation, http://www.mountainlion.org/featurevideo/featurevideojaybruce.asp, accessed July 18, 2016.

11. John Schroeder, "Giles Goswick—Pioneer Lion Hunter," *Prescott Evening Courier,* July 14, 1966, https://news.google.com/newspapers?nid=897&dat=19660714&id

=S6pMAAAAIBAJ&sjid=aFADAAAAIBAJ&pg=4987,3412392&hl=en, accessed July 7, 2015.

12. Mark Elbroch, in discussion with the author, July 19, 2015.

13. "F51–Honoring the Life of a Female Cougar," *Panthera,* last updated April 17, 2014, http://www.panthera.org/node/4768, accessed August 7, 2015; ibid.

14. Mark Elbroch, in discussion with the author, July 19, 2015.

15. Ibid.

16. Ibid.

17. Guy A. Balme and Luke T. B. Hunter, "Why Leopards Commit Infanticide," *Animal Behaviour* 86, no. 4 (2013): 791–799.

18. Craig B. Stanford, *Significant Others: The Ape-Human Continuum and the Quest for Human Nature* (New York: Basic Books, 2001), 77.

19. Anne Innis Dagg, "Infanticide by Male Lions Hypothesis: A Fallacy Influencing Research into Human Behavior," *American Anthropologist* 100, no. 4 (1998): 940–950.

20. Craig Packer, "Infanticide Is No Fallacy," *American Anthropologist* 102, no. 4 (2000): 829.

21. R. Brian Ferguson, "Born to Live: Challenging Killer Myths," in Robert W. Sussman and C. Robert Cloninger, *Origins of Altruism and Cooperation* (New York: Springer, 2011), 249–270; http://blogs.scientificamerican.com/cross-check/war-scholar-critiques-new-study-of-roots-of-violence/.

22. G. A. Bradshaw, Theodora Capaldo, Lorin Lindner, and Gloria Grow, "Developmental Context Effects on Bicultural Post-Trauma Self-Repair in Chimpanzees," *Developmental Psychology* 45, no. 5 (2009): 1376.

23. J. David Sweatt, "The Emerging Field of Neuroepigenetics," *Neuron* 80, no. 3 (2013): 624.

24. Dagg, "Infanticide," 948.

25. Ibid.

26. G. A. Bradshaw and Allan N. Schore, "How Elephants Are Opening Doors: Developmental Neuroethology, Attachment and Social Context," *Ethology* 113, no. 5 (2007): 426–436.

27. L. Mark Elbroch and Howard Quigley, "Social Interactions in a Solitary Carnivore," *Current Zoology,* (in press); Mark Elbroch, in discussion with the author, August 18, 2015.

28. Harley Shaw, in discussion with the author, July 19, 2015.

29. John Scott, *Social Network Analysis* (Thousand Oaks, CA: Sage, 2012).

30. Darcia Narvaez, *Embodied Morality: Protectionism, Engagement and Imagination,* (New York: Palgrave MacMillan, 2016).

31. Darcia Narvaez, *Neurobiology and the Development of Human Morality: Evolution, Culture, and Wisdom* (New York: W.W. Norton, 2014), 248.

32. Mark Elbroch, "Mountain Lion Dispersal," *National Geographic (blogs),* last

updated April 13, 2015, http://voices.nationalgeographic.com/2015/04/13/mountain-lion-dispersal, accessed August 30, 2015.

33. Ibid.

34. Mark Elbroch in film clip "Hippy Mother Mountain Lion," http://www.youtube.com/watch?v=Mu8Xh8CFZ5k, accessed August 30, 2015.

35. Elbroch, "Mountain Lion Dispersal."

36. "Family of Mountain Lions Spotted on UC Berkeley Campus," last updated July 24, 2012, http://sanfrancisco.cbslocal.com/2012/07/24/family-of-mountain-lions-spotted-on-uc-berkeley-campus, accessed August 30, 2015.

37. "BBC News, Hollywood Hills Mountain Lion Captured on Film," last updated Dec 4, 2013, http://www.youtube.com/watch?v=tHJkcqSwNjs, accessed August 30, 2015.

38. Adrian Glick Kudler, "Griffith Park Mountain Lion Twitters That He Ate Tom LaBonge," *Curbed LA,* last updated August 15, 2012, http://la.curbed.com/archives/2012/08/griffith_park_mountain_lion_twitters_that_he_ate_tom_labonge.php, accessed August 30, 2015.

39. Seth P. D. Riley, Laurel E. K. Serieys, and Joanne G Moriarty, "Infectious Disease and Contaminants in Urban Wildlife: Unseen and Often Overlooked Threats," in Robert A. McCleery, Christopher E. Moorman, and M. Nils Peterson, eds., *Urban Wildlife Conservation* (New York: Springer, 2014), 175–215.

40. Laurel Klein Serieys, "About," *Urban Carnivores,* http://www.urbancarnivores.com/about-us, accessed August 30, 2015.

41. "Mountain Lions of L.A.," *Sixty Minutes,* http://www.cbsnews.com/news/60-minutes-mountain-lions-of-los-angeles, accessed February 12, 2016.

42. Catherine M. S. Lambert et al., "Cougar Population Dynamics and Viability in the Pacific Northwest," *Journal of Wildlife Management* 70, no. 1 (2006): 246–254.

43. Harley Shaw, in discussion with the author, March 4, 2015.

44. Allan N. Schore, *Affect Dysregulation and Disorders of the Self* (New York: Norton, 2003); Narvaez, *Neurobiology.*

45. Virginia Hughes, "Epigenetics: The Sins of the Father," *Nature News and Comment,* last modified March 5, 2014, http://www.nature.com/news/epigenetics-the-sins-of-the-father-1.14816, accessed August 30, 2015.

46. Ole-Gunnar Støen et al., "Physiological Evidence for a Human-Induced Landscape of Fear in Brown Bears (*Ursus arctos*)," *Physiology and Behavior* 152 (2015): 244–248; M. A. Ditmer et al., "Behavioral and Physiological Responses of American Black Bears to Landscape Features within an Agricultural Region," *Ecosphere* 6, no. 3 (2015): 1–21.

47. B. B. Ackerman, F. G. Lindzey, and T. P. Hemker, "Cougar Food Habits in Southern Utah," *Journal of Wildlife Management* 48 (1984): 147–155.

48. L. Mark Elbroch et al., "The Difference between Killing and Eating: Ecological Shortcomings of Puma Energetic Models," *Ecosphere* 5, no. 5 (2014): art53, http://dx.doi.org/10.1890/ES13-00373.1, accessed June 27, 2016.

49. Bianca Barragan, "Meet the Murderous, Inbred Mountain Lion Family of the Santa Monica Mountains," *Curbed LA,* last updated March 4, 2015, http://la.curbed.com/archives/2015/03/meet_the_murderous_inbred_mountain_lion_family_of_the_santa_monica_mountains.php, accessed August 30, 2015.

50. "Lions of LA," http://www.lionslivehere.com, accessed August 30, 2015.

51. G. A. Bradshaw et al., "Elephant Breakdown," *Nature* 433, no. 7028 (2005): 807.

52. Ibid.; Bradshaw and Schore, "How Elephants Are Opening Doors," 426–436.

53. Steve P. Galentine and Pamela K. Swift, "Intraspecific Killing among Mountain Lions (*Puma concolor*)," *Southwestern Naturalist* 52, no. 1 (2007): 168.

54. Bradshaw, "Elephant Breakdown"; Gay A. Bradshaw et al., "Developmental Context Effects on Bicultural Posttrauma Self Repair in Chimpanzees," *Developmental Psychology* 45, no. 5 (2009): 1376–1388, http://www.ncbi.nlm.nih.gov/pubmed/19702399, accessed June 27, 2016; G. A Bradshaw et al., "Building an Inner Sanctuary: Complex PTSD in Chimpanzees," *Journal of Trauma and Dissociation* 9, no. 1 (2008): 9–34; G. A. Bradshaw and R. M. Sapolsky, "Mirror, Mirror," *American Scientist* 94, no. 6 (2006): 487–489, http://www.americanscientist.org/issues/pub/mirror-mirror-1, accessed June 27, 2016.

55. Bradshaw, *Elephants on the Edge;* Jay S. Mallonée and Paul Joslin, "Traumatic Stress Disorder Observed in an Adult Wild Captive Wolf (*Canis lupus*)," *Journal of Applied Animal Welfare Science,* 7(2), 107–26.

56. Narvaez, *Neurobiology,* 5.

57. Ibid.; D. Narvaez and K. Mrkva, "Creative Moral Imagination," in Seana Moran, David Cropley, and James C. Kaufman, eds., *The Ethics of Creativity* (New York: Palgrave Macmillan, 2014), 25–45.

Chapter 7. Coyotes

1. Indiana Coyote Rescue Center, Home Page, http://www.coyoterescue.org, accessed August 30, 2015.

2. Gay A. Bradshaw, *Elephants on the Edge: What Animals Teach Us about Humanity* (New Haven: Yale University Press, 2009); Theodora Capaldo and G. A. Bradshaw, *The Bioethics of Great Ape Well-Being: Psychiatric Injury and Duty of Care* (Ann Arbor, MI: Animals and Society Institute, 2011); G. A. Bradshaw, J. Yenkosky, and E. McCarthy, "Avian Affective Dysregulation: Psychiatric Models and Treatment for Parrots in Captivity," *Proceedings of the Association of Avian Veterinarians, 28th Annual Conference, Minnesota, 2009;* Jessica Bell Rizzolo and G. A. Bradshaw, "Complex Post Traumatic Stress Disorder in Asian Elephants Subjected to Social Deprivation and Breaking (*Phajaan*)," *Journal of Applied Animal Welfare Science* (in press).

3. Horacio Fabrega, "Making Sense of Behavioral Irregularities of Great Apes," *Neuroscience and Biobehavioral Reviews* 30, no. 8 (2006): 1260–1273; G. A. Bradshaw et al., "Developmental Context Effects on Bicultural Post-Trauma Self Repair in Chimpanzees," *Developmental Psychology* 45, no. 5 (2009): 1380, http://www.ncbi.nlm.nih.gov/pubmed/19702399, accessed June 27, 2016, doi:10.1037/a0015860.

4. Jami Hammer, in discussion with the author, July 2015.

5. Emma R. Bush, Sandra E. Baker, and David W. Macdonald, "Global Trade in Exotic Pets 200–12," *Conservation Biology* 28, no. 3 (2014): 663–676.

6. Sandra E. Baker et al., "Rough Trade: Animal Welfare in the Global Wildlife Trade," *BioScience* 63, no. 12 (2013): 928–938; Samantha Novic, "Exotic Pets Jam Arizona Wildlife Sanctuaries," *East Valley Tribune,* last updated May 7, 2007, http://www.eastvalleytribune.com/arizona/article_7ca85357-880d-5017-961a-36f687721027.html, accessed August 30, 2015.

7. Allan M. Casey III and Shirley J. Casey, "A Survey and Study of State Regulations Governing Wildlife Rehabilitation—2004 Update: Current Regulations and 10-Year Trends," *Journal of Wildlife Rehabilitation* 23 (2005): 17–85.

8. Julie Zauzmer, "A Man Asked Police for Help Finding His Illegal Pet. Now It's Dead, and He Could Be Arrested," *Washington Post,* http://www.washingtonpost.com/local/crime/man-might-be-arrested-for-owning-losing-coyote/2015/08/27/b400d65e-4ce2–11e5-bfb9–9736d04fc8e4_story.html, accessed August 30, 2015.

9. Jami Hammer, in discussion with the author, July 2015.

10. Ibid.

11. "Importation, Transportation and Possession of Live Restricted Animals," *Animal Legal and Historical Center,* last updated May 2015, https://www.animallaw.info/administrative/california-exotic-pets-importation-transportation-and-possession-live, accessed August 30, 2015.

12. G. A. Bradshaw and M. Engebretson, "Parrot Breeding and Keeping: Impacts of Capture and Captivity on Bird Wellbeing" (Ann Arbor, MI: Animals and Society Institute, 2013), also in Spanish; G. A. Bradshaw and Allan N. Schore, "How Elephants Are Opening Doors: Developmental Neuroethology, Attachment and Social Context," *Ethology* 113, no. 5 (2007): 426–436; Bradshaw et al., "Developmental Context Effects"; G. A. Bradshaw et al., "Building an Inner Sanctuary: Trauma-Induced Symptoms in Non-Human Great Apes," *Journal of Trauma and Dissociation* 9, no. 1 (2008): 9–34.

13. Carol Buckley, in discussion with the author, January 2011.

14. Bradshaw and Engebretson, "Parrot Breeding and Keeping"; Bradshaw et al., "Developmental Context Effects," 1376–1388; G. A. Bradshaw, "Elephants in Captivity: Analysis of Practice, Policy, and the Future," *Society and Animals* (2007): 1–48; Bradshaw et al., "Building an Inner Sanctuary."

15. Darlene D. Francis et al., "Environmental Enrichment Reverses the Effects of Maternal Separation on Stress Reactivity," *Journal of Neuroscience* 22, no. 18 (2002): 7840–7843.

16. Deborah Young et al., "Environmental Enrichment Inhibits Spontaneous Apoptosis, Prevents Seizures, and Is Neuroprotective," *Nature Medicine* 5, no. 4 (1999): 448–453.

17. Eleanor A. Maguire et al., "Navigation-Related Structural Change in the Hippocampi of Taxi Drivers," *Proceedings of the National Academy of Sciences* 97, no. 8 (2000): 4398–4403.

18. Deborah Blum, *The Monkey Wars* (Oxford, Eng.: Oxford University Press, 1994); Robert Mearns Yerkes, *Almost Human* (London: J. Cape, 1925).

19. Heini Hediger, *Studies of the Psychology and Behavior of Captive Animals in Zoos and Circuses* (N.p., 1955); Jill Mellen and Marty Sevenich MacPhee, "Philosophy of Environmental Enrichment: Past, Present, and Future," *Zoo Biology* 20, no. 3 (2001): 211–226; David J. Shepherdson, Jill D. Mellen, and Michael Hutchins, eds., *Second Nature: Environmental Enrichment for Captive Animals* (Washington, DC: Smithsonian Institution, 2012).

20. Jami Hammer, in discussion with the author, July 31, 2015.

21. Hope Ryden, *God's Dog: A Celebration of the North American Coyote* (Lincoln, NE: Backinprint.com, 2005), 27.

22. Kevin Hansen and Robert Redford, *Cougar: The American Lion* (Flagstaff, AZ: Northland Press, 1992), 56.

23. Tom Knudson, "The Killing Agency: Wildlife Services' Brutal Methods Leave a Trail of Animal Death," *Sacramento Bee,* April 28, 2012, http://www.sacbee.com/news/investigations/wildlife-investigation/article2574599.html, accessed February 2016.

24. Matthew E. Gompper, "The Ecology of Northeast Coyotes," *Wildlife Conservation Society* 17 (2002): 1–47.

25. Lee Hall, "Beyond a Government-the-Hunter Paradigm: Challenging Government Policies on Deer in a Critical Ecological Era," *Journal of Environmental Law and Litigation* 30 (2015): 265.

26. Steve Spaleta, "Coyote Shot for 'Sport' and Killed by Dogs," *LiveScience.com,* http://www.livescience.com/46313-coyote-shot-for-sport-and-killed-by-dogs-graphic-video.html, accessed August 31, 2015.

27. Jami Hammer, in discussion with the author, July 31, 2015.

28. Jami Hammer, in discussion with the author, August 17, 2015.

29. Ibid.

30. "TV Guide's 60 Nastiest Villains of All Time," last updated December 4, 2013, http://www.listal.com/list/tv-guides-60-nastiest-villains, accessed August 31, 2015.

31. Jami Hammer, in discussion with the author, July 31, 2015.

32. John Shivik, *The Predator Paradox: Ending the War with Wolves, Bears, Cougars, and Coyotes* (Boston: Beacon Press, 2014), 62.

33. Natural Resources Defense Council, "Reform Wildlife Services" http://www.nrdc.org/wildlife/animals/wolves/predatorcontrol.asp, accessed March 7, 2016; Hansen and Redford, *Cougar,* 59.

34. Shivik, *Predator Paradox,* 64.

35. Ibid.

36. William J. Ripple et al., "Widespread Mesopredator Effects after Wolf Extirpation," *Biological Conservation* 160 (2013): 70–79.

37. Mark Weckel et al., "Coyotes Go 'Bridge and Tunnel': A Narrow Opportunity to Study the Socio-Ecological Impacts of Coyote Range Expansion on Long Island, NY Pre-and Post-Arrival," *Cities and the Environment (CATE)* 8, no. 1 (2015): 5; Andrea S. Laliberte and William J. Ripple, "Range Contractions of North American Carnivores and Ungulates," *BioScience* 54, no. 2 (2004): 123–138.

38. J. Kirbey, "Why Coyotes Are Flourishing in New York City," *New York Magazine,* May 2015, http://nymag.com/daily/intelligencer/2015/05/coyotes-new-york-city.html#, accessed June, 2105.

39. John F. Benson et al., "Genetic and Environmental Influences on Pup Mortality Risk for Wolves and Coyotes within a Canis Hybrid Zone," *Biological Conservation* 166 (2013): 133–141.

40. Shivik, *Predator Paradox,* 64; U.S. Census 2011, http://www.census.gov/popest/data/historical/2010s/vintage_2011, accessed March 2016.

41. Wildlife Service, Division of Federal Aid and U.S. Bureau of the Census, *National Survey of Fishing, Hunting, and Wildlife-Associated Recreation* (2011; Washington, DC: Division of Federal Aid, 2014).

42. Stephen R. Kellert, "Public Perceptions of Predators, Particularly the Wolf and Coyote," *Biological Conservation* 31, no. 2 (1985): 167–189, esp. 168–169.

43. Ryden, *God's Dog,* 67.

44. Maxwell quoted in ibid., 131.

45. Shivik, *Predator Paradox,* 63.

46. Ibid.

47. Gross costs of $790,108,758.09 (hunting) and $637,650,071.38 (fishing). See U.S. Fish and Wildlife Service, "Historical Fishing License Data," http://wsfrprograms.fws.gov/Subpages/Licen+seInfo/Fishing.htm, accessed July 15, 2015 and U.S. Fish and Wildlife Service, "Historical Hunting License Data," http://wsfrprograms.fws.gov/Subpages/LicenseInfo/Hunting.htm, accessed July 15, 2015.

48. National Shooting Sports Foundation, "Hunting in America: An Economic Force for Conservation," 2013, accessed July 19, 2016; International Association of

Fish and Wildlife Agencies, "Economic Importance of Hunting in America," 2002, http://digitalcommons.law.msu.edu/afwa, accessed July 2016.

49. "Global Hunting Industry Reflects Economy," Reuters, January 11, 2010, http://www.reuters.com/article/2010/01/11/us-safari-industry-recession-idUSTRE60A3HN20100111, accessed June 26, 2016.

50. "Can Trophy Hunting Actually Help Conservation?" *Conservation,* January 15, 2014, http://conservationmagazine.org/2014/01/can-trophy-hunting-reconciled-conservation, accessed June 26, 2016; *The $200 Million Question: How Much Does Trophy Hunting Really Contribute to African Communities?,* a report for the African Lion Coalition prepared by Economists at Large, Melbourne, Australia, 2013, http://www.ecolarge.com/our-work, accessed July 17, 2016.

51. Quoted in "Endangered: What Is, What Isn't, Who Says," *Range* (Fall 2002): 5.

52. Shivik, *Predator Complex,* 10.

53. Ibid., 10–11.

54. Hansen and Redford, *Cougar,* 57.

55. USDA Animal and Plant Health Inspection Service, "Wildlife Damage," http://www.aphis.usda.gov/wps/portal/aphis/ourfocus/wildlifedamage, accessed August 31, 2015.

56. Shivik, *Predator Paradox,* 12.

57. U.S. Fish and Wildlife Service, "Gray Wolves Delisted in Western Great Lakes Distinct Population Segment," http://www.fws.gov/midwest/Wolf/delisting/index.htm, accessed August 31, 2015.

58. M. Gannon, "Unsettled Science behind Proposal to Lift Gray Wolf Protections, Panel Says," *LiveScience,* February 7, 2007, http://www.livescience.com/43213-unsettled-science-gray-wolf-peer-review.html, accessed November 2015.

59. James A. Foley, "U.S. Used Unsound Science in Making Recommendations to Delist Gray Wolf from Endangered Species List," *Nature World News* (2014), http://www.natureworldnews.com/articles/5935/20140207/used-unsound-science-making-recommendations-delist-gray-wolf-endangered-species.htm, accessed November 2015.

60. "Federal Court Puts Gray Wolf Back on Endangered Species List," CBS News, last updated December 19, 2014, http://www.cbsnews.com/news/federal-court-rejects-obama-decision-to-delist-great-lakes-wolf-popuation, accessed August 31, 2015; ibid.; Gannon, "Unsettled Science"; Foley, "U.S. Used Unsound Science."

61. Shire quoted in Virginia Morrell, "Judge Returns Great Lakes Wolves to Endangered Species List," December 2014, http://www.sciencemag.org/news/2014/12/judge-returns-great-lakes-wolves-endangered-species-list, accessed August, 2015.

62. Liz Klebaner, "Third Gray Wolf Delisting Bill Introduced in House," *Endangered Species Law and Policy,* last updated April 27, 2015, http://www.endangeredspecies

lawandpolicy.com/2015/04/articles/delisting/third-gray-wolf-delisting-bill-introduced-in-house, accessed August 31, 2015.

63. Jimenez quoted in Virginia Morell, "More Radio-Collared Wolves in Wyoming Shot Dead," *Science Magazine,* last updated December 17, 2012, http://news.sciencemag.org/people-events/2012/12/more-radio-collared-wolves-wyoming-shot-dead, accessed August 31, 2015.

64. Aaron Kunz, "Shooting Wolves in a Barrel—Radio Collar Info Given to Hunters," *Wolf Web,* last updated June 2, 2011, http://wolfweb.com/?p=101. http://wolfweb.com/?p=101, accessed July 17, 2015.

65. Oregon Department of Fish and Wildlife, "Wolves in Oregon," http://www.dfw.state.or.us/wolves, accessed August 15, 2015.

66. C. J. Baker, "Judge Rules for Government in Bear Mauling Suit," last modified October 25, 2012, http://www.powelltribune.com/news/item/10271-judge-rules-for-government-in-bear-mauling-suit, accessed August 31, 2015.

67. "Dayton Halts Moose Radio Collaring, But Scientists Outraged," *TwinCities.com,* last updated April 29, 2015, http://www.twincities.com/localnews/ci_28008482/dayton-halts-moose-radio-collaring-but-scientists-outraged, accessed August 4, 2015.

68. Ibid.

69. Stephen Rex Brown, "Moose on the Loose! Cantankerous Calf Attacks Biologist in Maine," *NY Daily News,* last updated February 21, 2014, http://www.nydailynews.com/news/national/video-moose-loose-cantankerous-calf-attacks-biologist-maine-article-1.1622458, accessed August 4, 2015.

70. L. Mark Elbroch et al., "Trailing Hounds vs. Foot Snares: Comparing Injuries to Pumas (*Puma concolor*) Captured in Chilean Patagonia," *Wildlife Biology* 19, no. 2 (2013): 210.

71. Ibid., 214.

72. "Tired Mountain Lion," *YouTube,* http://www.youtube.com/watch?v=ynTswMPSUe4, accessed August 4, 2015.

73. Brendan Bane, "How to Capture a Mountain Lion," *Hilltromper Santa Cruz,* http://santacruz.hilltromper.com/article/scientists-hounds-capture-mountain-lions, accessed August 4, 2015.

74. Mark A. Ditmer et al., "Bears Show a Physiological but Limited Behavioral Response to Unmanned Aerial Vehicles," *Current Biology* 25, no. 17 (2015), http://linkinghub.elsevier.com/retrieve/pii/S0960982215008271, accessed August 21, 2015.

75. Bruce D. Perry, "Neurobiological Sequelae of Childhood Trauma: PTSD in Children," in M. Murburg, ed., *Catecholamine Function in Post Traumatic Stress Disorder: Emerging Concepts* (Washington, DC: American Psychiatric Press, 1994), 253–276; Eberhard von Borell et al., "Heart Rate Variability as a Measure of Autonomic

Regulation of Cardiac Activity for Assessing Stress and Welfare in Farm Animals—A Review," *Physiology and Behavior* 92, no. 3 (2007): 293–316.

76. Mallika Rao, "Here's Why Walter Palmer Keeps Saying He 'Took' Cecil The Lion," *Huffington Post,* last updated August 8, 2015, http://www.huffingtonpost.com/entry/walter-palmer-lion-take-kill-word-choice_55b8e465e4b0a13f9d1aea2c, accessed August 10, 2015.

77. Ibid.

78. Lee Burkins, *Soldier's Heart: An Inspirational Memoir and Inquiry of War* (Bloomington, IN: 1st Books, 2002), 255.

79. U.S. Fish and Wildlife Service, "NCDE Grizzly Bear Conservation Strategy" (April 2013): ii, http://www.fws.gov/mountain-prairie/science/PeerReviewDocs/NCDE_Grizzly.pdf, accessed August 4, 2015.

80. Ibid., 122–123.

81. Peter Verney, *Animals in Peril: Man's War against Wildlife* (Provo, UT: Brigham Young University Press, 1979).

82. U.S. Fish and Wildlife Service, "Mission Statement, National Wildlife Refuge System," http://www.fws.gov/refuges/about/mission.html, accessed July 15, 2015.

83. Ibid.

84. Anonymous, in discussion with the author, October 2014.

85. Jacqueline Ingles, "Mary Musselman, 81, Died Friday; Back in January She Was Jailed for a," *WFTS,* last updated August 30, 2014, http://www.abcactionnews.com/news/region-tampa/mary-musselman-81-died-friday-back-in-january-she-was-jailed-for-a, accessed August 1, 2015.

86. Robbie Feinberg, "How Bear Expert Lynn Rogers Went from Scientific Pioneer to Pariah," *City Pages,* last updated August 30, 2014, http://www.citypages.com/news/how-bear-expert-lynn-rogers-went-from-scientific-pioneer-to-pariah-6780931, accessed August 24, 2015.

87. "Conservation Officers Allegedly Catch Suspected Marijuana Farmer of Feeding Black Bears Again," *National Post,* last updated November 16, 2011, http://news.nationalpost.com/news/canada/conservation-officers-allegedly-catch-suspected-marijuana-farmer-of-feeding-black-bears-again, accessed August 31, 2015.

88. Ibid.

89. Mat-Su Convention and Visitors Bureau, Home Page, http://www.alaskavisit.com, accessed August 31, 2015.

90. Alaska Dispatch News, "'Bear Haven' Owner Charlie Vandergaw at Sentencing," http://www.youtube.com/watch?v=E_9cAOTEakE, accessed July 1, 2015.

91. Alaska Department of Fish and Game, "Living in Harmony with Bears: Bears and 'Food,'" http://www.adfg.alaska.gov/index.cfm?adfg=livingwithbears.bearharmony, accessed August 3, 2015.

92. Michigan Department of Natural Resources, "What Is the Difference between

Supplemental Feeding and Baiting?," http://www.michigan.gov/dnr/0,4570,7-153-10366_37141_37705-31604-,00.html, accessed June 30, 2015.

93. Joseph von Benedikt and Skip Knowles, "Spot-and-Stalk vs. Baiting: What's the Best Bear Hunting Method?," *Petersen's Hunting,* last updated April 18, 2014, http://www.petersenshunting.com/gear-accessories/versus/spot-stalk-vs-baiting-whats-best-bear-hunting-method, accessed August 31, 2015.

94. "Bear Baiting: Tips" *Alaska Department of Fish and Game,* http://www.adfg.alaska.gov/index.cfm?adfg=bearbaiting.tips, accessed July 20, 2015.

95. Alaska Department of Fish and Game. "Living in Harmony with Bears."

96. Ibid.

97. Darryl Fears, "The Killing Fields: USDA Says It Slaughtered More Than 4 Million Wild Animals Last Year," *Washington Post,* June 6, 2014, http://www.washingtonpost.com/national/health-science/governments-kill-of-4-million-animals-seen-as-anoverstep/2014/06/06/1de0c550-ecc4-11e3-b98c-72cef4a00499_story.html, accessed August 31, 2015.

98. Nancy Macdonald, "Grizzly Toll: British Columbia's Controversial Trophy Bear Hunt," *Macleans.ca,* last updated October 24, 2014, http://www.macleans.ca/society/grizzly-toll-b-cs-controversial-trophy-bear-hunt, accessed August 31, 2015.

99. Darryl Fears, "The Killing Fields: USDA Says It Slaughtered More Than 4 Million Wild Animals Last Year," *Washington Post,* June 6, 2014, http://www.washingtonpost.com/national/health-science/governments-kill-of-4-million-animals-seen-as-anoverstep/2014/06/06/1de0c550-ecc4-11e3-b98c-72cef4a00499_story.html, accessed August 31, 2015.

100. "Torturing Statistics: Wildlife Services' Big Carnivore Kill Rate Is 98%," *Melissa Waage's Switchboard Blog (NRDC),* last updated July 11, 2012, http://switchboard.nrdc.org/blogs/mwaage/torturing_statistics_wildlife.html, accessed July 30, 2015.

101. "Bill Would Prevent Wildlife Services and Others from Cruel Trapping on Refuge Land," *Elly Pepper's Switchboard Blog (NRDC),* last updated April 27, 2015, http://switchboard.nrdc.org/blogs/epepper/bill_would_prevent_wildlife_se.html, accessed August 1, 2015.

102. USDA Animal and Plant Health Inspection Service, "Wildlife Damage Management," http://www.aphis.usda.gov/wildlife_damage/prog_data/2011_prog_data/index.shtml, accessed August 3, 2015.

103. James Gorman, "Mass Animal Deaths: An Environmental Whodunit," *New York Times,* last updated January 8, 2011, http://www.nytimes.com/2011/01/09/weekinreview/09gorman.html, accessed July 31, 2015.

104. "Blackbird Deaths in Beebe Intentional," *www.Fox16.com,* last updated January 2, 2012, http://www.fox16.com/news/story/Blackbird-deaths-in-Beebe-intentional/d/story/wOBjGOehWE-ZyIi-PR5qxw, accessed August 1, 2015; USDA Animal and Plant Health Inspection Service, Wildlife Services, *Reducing Blackbird Damage to Sprout-*

ing Rice through an Integrated Wildlife Damage Management Program in Southwestern Louisiana (Washington, DC: USDA, 2001), http://www.aphis.usda.gov/regulations/pdfs/nepa/LA_Blackbird_EA.pdf, also https://webcache.googleusercontent.com/search?q=cache:Thg7MQxFPXYJ:https://www.aphis.usda.gov/regulations/pdfs/nepa/LA%2520Blackbird%2520EA.pdf+&cd=1&hl=en&ct=clnk&gl=us, accessed June 4, 2015.

105. Ibid.

106. Oregon Department of Fish and Wildlife, "Mentored Youth," http://www.dfw.state.or.us/education/mentored_youth, accessed August 7, 2015.

107. Macdonald, "Grizzly Toll."

108. Elaine O'Flynn, "Paul McCartney Says David Cameron Will Lose Support over Fox Hunting," *Daily Mail Online,* last updated July 11, 2015, http://www.dailymail.co.uk/news/article-3157384/Sir-Paul-McCartney-warns-David-Cameron-lose-support-British-people-fox-hunting-reintroduced.html, accessed August 6, 2015.

109. Mina Abgoon, "Major Support for Conservation Officer Suspended for Refusing to Kill Bear Cubs," *NowTrending.com,* last updated July 11, 2015, http://www.scrippsmedia.com/now-trending/Major-support-for-conservation-officer-suspended-for-refusing-to-kill-bear-cubs-313897751.html, accessed August 31, 2015.

110. "Stranded Shark Rescued from Massachusetts Beach," *BBC News,* last updated July 20, 2015, http://www.bbc.com/news/world-us-canada-33533620, accessed August 31, 2015.

111. Sara Gates, "Man Brutally Beat Great White Shark to Death," *Huffington Post,* February 13, 2014, http://www.huffingtonpost.com/2014/02/13/beating-great-white-shark-death-man-fined-australia_n_4781644.html, accessed June 5, 2015.

112. "Encyclical Letter *Laudato Si'* of The Holy Father Francis on Care for Our Common Home," *The Holy See,* May 24, 2015, http://w2.vatican.va/content/francesco/en/encyclicals/documents/papa-francesco_20150524_enciclica-laudato-si.html, accessed August 31, 2015.

113. Sergio Cisneros, "Oregon Fish and Wildlife Face $32 Million Budget Shortfall," *OPB,* last modified June 6, 2014, http://www.opb.org/news/article/oregon-fish-and-wildlife-face-32-million-budget-shortfall, accessed August 31, 2015.

114. Oregon Department of Fish and Wildlife, "ODFW Budget Information," http://www.dfw.state.or.us/agency/budget, accessed August 31, 2015.

115. Idaho Fish and Game Department, "Idaho Hunt Planner," http://fishandgame.idaho.gov/ifwis/huntPlanner, accessed August 31, 2015.

116. Colorado Parks and Wildlife, "Hunting Statistics," http://cpw.state.co.us/thingstodo/Pages/Statistics.aspx, accessed August 31, 2015.

117. Hunting, once the purview of men, now involves a burgeoning number of women. The Women Hunters organization, with its tagline "For women, about women,

by women" is "dedicated to the encouragement, education and promotion of women and youth in the hunting traditions . . . Women hold the key to the future of hunting and WomenHunters™ will contribute to the positive growth of hunting for future generations."

118. Oregon Department of Fish and Wildlife, "Facts about the Mentored Youth Hunter Program," http://www.dfw.state.or.us/education/mentored_youth/facts.asp, accessed August 31, 2015.

119. Oregon Department of Fish and Wildlife, "Mentored Youth," http://www.dfw.state.or.us/education/mentored_youth, accessed August 31, 2015.

120. Lorraine Donlon, in discussion with the author, May 2, 2104.

121. G. A. Bradshaw, "Run Rabbit Run," *Psychology Today,* http://www.psychologytoday.com/blog/bear-in-mind/201307/run-rabbit-run, accessed August 11, 2015; Animals and Society Institute, "AniCare," http://www.animalsandsociety.org/pages/anicare, accessed August 3, 2015; Animal Protection of New Mexico, Home Page, http://www.apnm.org, accessed August 31, 2015; Animal Protection of New Mexico, "What Are the Implications of the Link?," http://apnm.org/publications/link_overview.pdf, accessed January 27, 2013.

122. Radley Balko, "Five Myths about America's Police," *Washington Post,* last updated December 5, 2014, https://www.washingtonpost.com/opinions/five-myths-about-americas-police/2014/12/05/35b1af44-7bcd-11e4-9a27-6fdbc612bff8_story.html, accessed August 31, 2015.

123. Anonymous, in discussion with the author.

124. "10 Best Deer Guns for Hunting Today," *Outdoor Life,* http://www.outdoorlife.com/blogs/hunting/10-best-deer-guns-hunting-today, accessed August 31, 2015.

125. Ryden, *God's Dog,* 27.

126. Ibid.

127. Ibid., 29.

128. P. Nabokov, *Native American Testimony: A Chronicle of Indian-White Relations from Prophesy to the Present* (London: Penguin Press, 1991).

129. U.N. General Assembly Resolution 260A (III) article 2, December 9, 1948.

130. Maria Yellow Horse Brave Heart and Lemyra M. DeBruyn, "The American Indian Holocaust: Healing Historical Unresolved Grief," *American Indian and Alaska Native Mental Health Research* 8, no. 2 (1998): 60–82.

131. "Shark Bites Swimmer on N.C.'s Outer Banks," *USA Today,* June 26, 2015, http://www.usatoday.com/story/news/nation/2015/06/26/shark-bites-swimmer-ncs-outer-banks/29363837, accessed August 3, 2015.

132. "'Carnivore Cleansing' Is Damaging Ecosystems, Scientists Warn," *Guardian,* January 9, 2014, http://www.theguardian.com/environment/2014/jan/09/carnivore-cleansing-damaging-ecosystems, accessed June 27, 2016; W. J. Ripple et al., "Status

and Ecological Effects of the World's Largest Carnivores," *Science* 343, no. 6167 (January 10, 2014): 6164, http://www.sciencemag.org/content/343/6167/1241484, accessed June 27, 2016.

133. The quotation is from U.S. National Park Service, "Managing Human-Bear Interactions," http://www.nps.gov/subjects/bears/humanbearinteractions.htm, accessed August 31, 2015.

134. Jennifer M. Germano et al., "Mitigation-Driven Translocations: Are We Moving Wildlife in the Right Direction?," *Frontiers in Ecology and the Environment* 13, no. 2 (2015): 100–105; Ben A. Minteer and James P. Collins, "Move It or Lose It? The Ecological Ethics of Relocating Species under Climate Change," *Ecological Applications* 20, no. 7 (2010): 1801–1804.

135. Jami Hammer, in discussion with the author, August 2015.

Epilogue

1. Fred Buyle, in discussion with the author, August 2015.

2. Ibid.

3. "The Warrior Within—A Tai Chi Documentary with Lee Burkins," *Energy Arts,* last updated April 3, 2015, http://www.energyarts.com/blog/bruce-frantzis/warrior-within-tai-chi-documentary-lee-burkins, accessed August 20, 2115.

Acknowledgments

I would first like to thank the individuals whom I interviewed for their amazing insights, courage, and passion. It is as much their book as it is mine. For her elegant, eloquent guidance and erudition, my literary agent, Anne Borchardt; for her brilliance and love of nature, my editor, Jean Thomson Black; and for their meticulous, conscientious creativity, and dedication, Julie Carlson and Margaret Otzel—my deepest gratitude. To Barb and Patrick Murray, Frog Crossing Foundation, Summerlee Foundation, Frank Davis, National Center for Ecological Analysis and Synthesis, Paige Turner, and Dr. Zohara Hieronimus, my thanks for the support that made this work possible. I would also like to thank my friends and colleagues—old and new—for their care, patience, and incredible dedication. Finally, my deepest appreciation for my finned, furred, feathered, and scaled family and community who have never shown anything but love, an enduring confidence, and commitment to peace and goodwill. Many will have passed by the time of this publication—victims of what is the subject of this book. It is my hope that this work will be of use for them so that their children and their children's children may flourish in joy and freedom for evermore.

Index

Index